EARTH • AT • RISK

Acid Rain
Alternative Sources of Energy
Animal Welfare
The Automobile and the Environment
Clean Air
Clean Water
Degradation of the Land
Economics of Environmental Protection
Environmental Action Groups
The Environment and the Law
Environmental Disasters
Extinction
The Fragile Earth
Global Warming
The Living Ocean
Nuclear Energy • Nuclear Waste
Overpopulation
The Ozone Layer
Recycling
The Rainforest
Solar Energy
Toxic Materials
What You Can Do for the Environment
Wilderness Preservation

EARTH • AT • RISK

GLOBAL WARMING

by Burkhard Bilger

Introduction by
Russell E. Train

Chairman of
the Board of Directors,
World Wildlife Fund and
The Conservation Foundation

CHELSEA HOUSE PUBLISHERS

new york philadelphia

CHELSEA HOUSE PUBLISHERS
EDITOR-IN-CHIEF: Remmel Nunn
MANAGING EDITOR: Karyn Gullen Browne
COPY CHIEF: Mark Rifkin
PICTURE EDITOR: Adrian G. Allen
ART DIRECTOR: Maria Epes
ASSISTANT ART DIRECTOR: Noreen Romano
MANUFACTURING MANAGER: Gerald Levine
SYSTEMS MANAGER: Lindsey Ottman
PRODUCTION MANAGER: Joseph Romano
PRODUCTION COORDINATOR: Marie Claire Cebrián

EARTH AT RISK
SENIOR EDITOR: Jake Goldberg
Staff for *Global Warming*
ASSOCIATE EDITOR: Karen Hammonds
SENIOR COPY EDITOR: Laurie Kahn
EDITORIAL ASSISTANT: Danielle Janusz
PICTURE RESEARCHER: Villette Harris
DESIGNER: Maria Epes
LAYOUT: Marjorie Zaum

5 7 9 8 6

Library of Congress Cataloging-in-Publication Data
Bilger, Burkhard.
Global Warming/by Burkhard Bilger; introduction by Russell E. Train.
p. cm.—(Earth at risk)
Includes bibliographical references and index.
Summary: Examines the phenomenon of global warming, discussing the greenhouse effect in its positive, life-giving configuration, and again as this mechanism is knocked out of balance by increased levels of carbon gases.
ISBN 0-7910-1575-0
0-7910-1625-0 (pbk.)
1. Global warming—Juvenile literature. 2. Greenhouse effect, Atmospheric—Juvenile literature. [1. Global warming. 2. Greenhouse effect, Atmospheric.] I. Title. II. Series. 91-17781
QC981.8.G56B46 1992 CIP
363.73'87—dc20 AC

CONTENTS

INTRODUCTION

Russell E. Train

Administrator, Environmental Protection Agency, 1973 to 1977; Chairman of the Board of Directors, World Wildlife Fund and The Conservation Foundation

There is a growing realization that human activities increasingly are threatening the health of the natural systems that make life possible on this planet. Humankind has the power to alter nature fundamentally, perhaps irreversibly.

This stark reality was dramatized in January 1989 when *Time* magazine named Earth the "Planet of the Year." In the same year, the Exxon *Valdez* disaster sparked public concern over the effects of human activity on vulnerable ecosystems when a thick blanket of crude oil coated the shores and wildlife of Prince William Sound in Alaska. And, no doubt, the 20th anniversary celebration of Earth Day in April 1990 renewed broad public interest in environmental issues still further. It is no accident then that many people are calling the years between 1990 and 2000 the "Decade of the Environment."

And this is not merely a case of media hype, for the 1990s will truly be a time when the people of the planet Earth learn the meaning of the phrase "everything is connected to everything else" in the natural and man-made systems that sustain our lives. This will be a period when more people will understand that burning a tree in Amazonia adversely affects the global atmosphere just as much as the exhaust from the cars that fill our streets and expressways.

Central to our understanding of environmental issues is the need to recognize the complexity of the problems we face and the relationships between environmental and other needs in our society. Global warming provides an instructive example. Controlling emissions of carbon dioxide, the principal greenhouse gas, will involve efforts to reduce the use of fossil fuels to generate electricity. Such a reduction will include energy conservation and the promotion of alternative energy sources, such as nuclear and solar power.

The automobile contributes significantly to the problem. We have the choice of switching to more energy efficient autos and, in the longer run, of choosing alternative automotive power systems and relying more on mass transit. This will require different patterns of land use and development, patterns that are less transportation and energy intensive.

In agriculture, rice paddies and cattle are major sources of greenhouse gases. Recent experiments suggest that universally used nitrogen fertilizers may inhibit the ability of natural soil organisms to take up methane, thus contributing tremendously to the atmospheric loading of that gas—one of the major culprits in the global warming scenario.

As one explores the various parameters of today's pressing environmental challenges, it is possible to identify some areas where we have made some progress. We have taken important steps to control gross pollution over the past two decades. What I find particularly encouraging is the growing environmental consciousness and activism by today's youth. In many communities across the country, young people are working together to take their environmental awareness out of the classroom and apply it to everyday problems. Successful recycling and tree-planting projects have been launched as a result of these budding environmentalists who have committed themselves to a cleaner environment. Citizen action, activated by youthful enthusiasm, was largely responsible for the fast-food industry's switch from rainforest to domestic beef, for pledges from important companies in the tuna industry to use fishing techniques that would not harm dolphins, and for the recent announcement by the McDonald's Corporation to phase out polystyrene "clam shell" hamburger containers.

Despite these successes, much remains to be done if we are to make ours a truly healthy environment. Even a short list of persistent issues includes problems such as acid rain, ground-level ozone and smog, and airborne toxins; groundwater protection and nonpoint sources of pollution, such as runoff from farms and city streets; wetlands protection; hazardous waste dumps; and solid waste disposal, waste minimization, and recycling.

Similarly, there is an unfinished agenda in the natural resources area: effective implementation of newly adopted management plans for national forests; strengthening the wildlife refuge system; national park management, including addressing the growing pressure of development on lands surrounding the parks; implementation of the Endangered Species Act; wildlife trade problems, such as that involving elephant ivory; and ensuring adequate sustained funding for these efforts at all levels of government. All of these issues are before us today; most will continue in one form or another through the year 2000.

Each of these challenges to environmental quality and our health requires a response that recognizes the complex nature of the problem. Narrowly conceived solutions will not achieve lasting results. Often it seems that when we grab hold of one part of the environmental balloon, an unsightly and threatening bulge appears somewhere else.

The higher environmental issues arise on the national agenda, the more important it is that we are armed with the best possible knowledge of the economic costs of undertaking particular environmental programs and the costs associated with not undertaking them. Our society is not blessed with unlimited resources, and tough choices are going to have to be made. These should be informed choices.

All too often, environmental objectives are seen as at cross-purposes with other considerations vital to our society. Thus, environmental protection is often viewed as being in conflict with economic growth, with energy needs, with agricultural productions, and so on. The time has come when environmental considerations must be fully integrated into every nation's priorities.

One area that merits full legislative attention is energy efficiency. The United States is one of the least energy efficient of all the industrialized nations. Japan, for example, uses far less energy per unit of gross national product than the United States does. Of course, a country as large as the United States requires large amounts of energy for transportation. However, there is still a substantial amount of excess energy used, and this excess constitutes waste. More fuel efficient autos and home heating systems would save millions of barrels of oil, or their equivalent, each year. And air pollutants, including greenhouse gases, could be significantly reduced by increased efficiency in industry.

I suspect that the environmental problem that comes closest to home for most of us is the problem of what to do with trash. All over the world, communities are wrestling with the problem of waste disposal. Landfill sites are rapidly filling to capacity. No one wants a trash and garbage dump near home. As William Ruckelshaus, former EPA administrator and now in the waste management business, puts it, "Everyone wants you to pick up the garbage and no one wants you to put it down!"

At the present time, solid waste programs emphasize the regulation of disposal, setting standards for landfills, and so forth. In the decade ahead, we must shift our emphasis from regulating waste disposal to an overall reduction in its volume. We must look at the entire waste stream, including product design and packaging. We must avoid creating waste in the first place. To the greatest extent possible, we should then recycle any waste that is produced. I believe that, while most of us enjoy our comfortable way of life and have no desire to change things, we also know in our hearts that our "disposable society" has allowed us to become pretty soft.

Land use is another domestic issue that might well attract legislative attention by the year 2000. All across the United States, communities are grappling with the problem of growth. All too often, growth imposes high costs on the environment—the pollution of aquifers; the destruction of wetlands; the crowding of shorelines; the loss of wildlife habitat; and the loss of those special places, such as a historic structure or area, that give a community a sense of identity. It is worth noting that growth is not only the product of economic development but of population movement. By the year 2010, for example, experts predict that 75% of all Americans will live within 50 miles of a coast.

It is important to keep in mind that we are all made vulnerable by environmental problems that cross international borders. Of course, the most critical global conservation problems are the destruction of tropical forests and the consequent loss of their biological capital. Some scientists have calculated extinction rates as high as 11 species per hour. All agree that the loss of species has never been greater than at the present time; not even the disappearance of the dinosaurs can compare to today's rate of extinction.

In addition to species extinctions, the loss of tropical forests may represent as much as 20% of the total carbon dioxide loadings to the atmosphere. Clearly, any international approach to the problem of global warming must include major efforts to stop the destruction of forests and to manage those that remain on a renewable basis. Debt for nature swaps, which the World Wildlife Fund has pioneered in Costa Rica, Ecuador, Madagascar, and the Philippines, provide a useful mechanism for promoting such conservation objectives.

Global environmental issues inevitably will become the principal focus in international relations. But the single overriding issue facing the world community today is how to achieve a sustainable balance between growing human populations and the earth's natural systems. If you travel as frequently as I do in the developing countries of Latin America, Africa, and Asia, it is hard to escape the reality that expanding human populations are seriously weakening the earth's resource base. Rampant deforestation, eroding soils, spreading deserts, loss of biological diversity, the destruction of fisheries, and polluted and degraded urban environments threaten to spread environmental impoverishment, particularly in the tropics, where human population growth is greatest.

It is important to recognize that environmental degradation and human poverty are closely linked. Impoverished people desperate for land on which to grow crops or graze cattle are destroying forests and overgrazing even more marginal land. These people become trapped in a vicious downward spiral. They have little choice but to continue to overexploit the weakened resources available to them. Continued abuse of these lands only diminishes their productivity. Throughout the developing world, alarming amounts of land rendered useless by overgrazing and poor agricultural practices have become virtual wastelands, yet human numbers continue to multiply in these areas.

From Bangladesh to Haiti, we are confronted with an increasing number of ecological basket cases. In the Philippines, a traditional focus of U.S. interest, environmental devastation is widespread as deforestation, soil erosion, and the destruction of coral reefs and fisheries combine with the highest population growth rate in Southeast Asia.

Controlling human population growth is the key factor in the environmental equation. World population is expected to at least double to about 11 billion before leveling off. Most of this growth will occur in the poorest nations of the developing world. I would hope that the United States will once again become a strong advocate of international efforts to promote family planning. Bringing human populations into a sustainable balance with their natural resource base must be a vital objective of U.S. foreign policy.

Foreign economic assistance, the program of the Agency for International Development (AID), can become a potentially powerful tool for arresting environmental deterioration in developing countries. People who profess to care about global environmental problems—the loss of biological diversity, the destruction of tropical forests, the greenhouse effect, the impoverishment of the marine environment, and so on—should be strong supporters of foreign aid planning and the principles of sustainable development urged by the World Commission on Environment and Development, the "Brundtland Commission."

If sustainability is to be the underlying element of overseas assistance programs, so too must it be a guiding principle in people's practices at home. Too often we think of sustainable development only in terms of the resources of other countries. We have much that we can and should be doing to promote long-term sustainability in our own resource management. The conflict over our own rain forests, the old growth forests of the Pacific Northwest, illustrates this point.

The decade ahead will be a time of great activity on the environmental front, both globally and domestically. I sincerely believe we will be tested as we have been only in times of war and during the Great Depression. We must set goals for the year 2000 that will challenge both the American people and the world community.

Despite the complexities ahead, I remain an optimist. I am confident that if we collectively commit ourselves to a clean, healthy environment we can surpass the achievements of the 1980s and meet the serious challenges that face us in the coming decades. I hope that today's students will recognize their significant role in and responsibility for bringing about change and will rise to the occasion to improve the quality of our global environment.

The clouds that perpetually surround the earth both cool the planet by blocking incoming sunlight and warm it by blocking outgoing heat radiation. The equilibrium thus established makes life possible.

chapter 1

A WARMER WORLD?

On July 19, 1988, the headline on the front page of the science section of the *New York Times* sounded the alarm: MAJOR "GREENHOUSE" IMPACT IS UNAVOIDABLE, EXPERTS SAY. The story went on to explain, "Momentum toward a global warming caused by the greenhouse effect is now so great that there is no way to avoid a significant rise in temperature in the coming century. . . . Even draconian measures to reduce the air pollution that is responsible for the global warming can only buy time to adjust to a warmer world and put a ceiling on the ultimate warming."

The rumors had been in the air for some time. Nine months earlier, the cover story of *Time* magazine carried the headline THE HEAT IS ON: CHEMICAL WASTES SPEWED INTO THE AIR THREATEN THE EARTH'S CLIMATE. And just a week before the *Times* article appeared, *Newsweek* had run a special report called "Heat Waves," which declared that "global mean temperatures may rise as much as 8 degrees in half a century, especially in the temperate and polar regions. The deserts of the Southwest will creep up on the Great Plains; sea levels will rise, threatening low-lying coasts."

Such reports, dramatic though they were, might have been easy to shrug off. But the *Times* story had two things going for it: scientific authority and timing. From news conferences to congressional hearings, atmospheric scientists were joining forces to share their concerns about rising temperatures with the public, and they were trying to get governments to do something about it.

The theory of global warming itself was not new. Since the time of Napoléon, scientists had known about the greenhouse effect—whereby gases such as carbon dioxide (CO_2) retain heat in the earth's atmosphere, much as the glass panes in a greenhouse let the sunlight in but keep its heat from escaping. As far back as 1896, Swedish chemist and Nobel laureate Svante Arrhenius had predicted that if the concentration of atmospheric CO_2—which industry spews into the air by the millions of metric tons every year—were to double, global temperatures would rise by between 9 and 11 degrees Fahrenheit (5 and 6 degrees Celsius).

Still, no one took global warming very seriously until 60 years later, when a young geochemist named Charles Keeling began monitoring the levels of CO_2 in the atmosphere from the top of a mountain in Hawaii. At the time of Keeling's first measurement, CO_2 had already risen by 10% since Arrhenius's predictions; by 1990 the quantity of CO_2 in the atmosphere had risen by 20%. By the middle of the next century, at the current rate of CO_2 production, CO_2 concentrations in the atmosphere will have doubled since Arrhenius's time.

Scientists had long reassured themselves that the world's oceans would absorb any excess CO_2 in the atmosphere via a simple chemical reaction whereby the CO_2 combines with calcium ions dissolved in the water and precipitates down to the ocean floor as calcium carbonate. However, in 1957, just as Keeling was beginning to measure CO_2 levels in the atmosphere, oceanographer Roger Revelle determined that the oceans could absorb only a limited amount of CO_2. With this safety cushion gone and Keeling's graph of CO_2 levels rising in a steep curve through the 1960s and into the 1970s, atmospheric scientists began in earnest to look into the question of global warming.

At universities and think tanks such as the National Center for Atmospheric Research (NCAR) in Washington, D.C., scientists began to study the intricately connected pieces in the puzzle of global climate: ocean currents, trade winds, the seasonal expansion and shrinking of the polar ice caps, the fluctuations in the earth's orbit, and so forth. By the mid-1970s, the first computer simulations of climate behavior, called *general circulation models*, or GCMs, had been devised, allowing researchers to predict how global temperatures and climate might change if levels of CO_2 continued to rise. Paleoclimatologists (historians of climate) began looking at fossil plants, ice cores, and the tracks left by ancient glaciers to determine how and why the earth's climate changed in the past and what this might explain about the future. Along the way it became clear that CO_2 was not the only gas that could cause temperatures to rise—gases such as methane, water vapor, nitrous oxide, and chlorofluorocarbons contributed to the greenhouse effect as well, and their concentrations in the atmosphere were also increasing.

Global warming and atmospheric research in general might have remained relatively obscure topics, well beyond the reach of newspaper headlines, if the climate itself had not lent a helping hand. The 1980s was the hottest decade in history and included 7 of the 10 hottest years ever recorded. By the summer of 1988, the hottest months in living memory throughout much of the Northern Hemisphere, the timing was perfect for a global warming panic. With more forest fires occurring than at any other time in history (including a particularly devastating one in Yellowstone National Park), droughts decimating millions of acres of crops in the wheat belts of North America and Asia, and freak storms devastating the coastlines of Asia and Mexico, the *Times* story about global warming was bound to meet with a receptive audience.

DEALING WITH DOUBT

The sudden interest in global warming since 1988 has offered atmospheric scientists a unique opportunity to publicize their work, but it has also presented them with a dilemma. Thinking purely as scientists, even those who staunchly believe in global warming have to admit that it is still only a theory: Predictions of future temperature increases are inconclusive, and the best computers still cannot foresee how rising temperatures will affect specific climates around the world. Thinking as concerned human beings, however, most scientists know that a global temperature rise of about 2° to 9°F (1° to 5°C) is a real possibility and that the public should be warned that such an increase could cause drought, famine, rising sea levels, and a host of other environmental disasters over the next 100 years.

While scientists are torn between maintaining their objectivity and addressing their concerns, their problems are compounded by politics and the media. In order to pass laws limiting CO_2 emissions and funding further climate research, politicians need straightforward explanations of global warming and unequivocal predictions of its effects, rather than the carefully worded statements made by most scientists. In order to write

Should significant global warming occur, glaciers such as this one at Russell Fiord in Alaska will melt and raise sea levels.

eye-catching headlines and dramatic stories, the media often simplify, distort, or even misrepresent those same carefully worded statements.

The most notorious example of this conflict between scientific accuracy and the need to dramatize a problem occurred on June 23, 1988, when atmospheric physicist James E. Hansen appeared before the Senate Energy Committee in Washington, D.C. Leader of a team of scientists who run a powerful computer climate model at the Goddard Space Flight Center in Maryland, Hansen has long believed that global warming is a great threat. Thinking it was time that scientists stopped "waffling so much" in their public statements, he told the senators and reporters in attendance that "the probability of a chance warming of [the magnitude of the last three decades] is about one percent. So, with 99 percent confidence we can state that the warming during this time period is a real warming trend."

Colorado senator Tim Wirth now had the strong scientific statement he needed to gain support for the bill he had written addressing global warming. The media had the dramatic headline it needed to grab the public's attention. Articles about global warming, such as the *Times* story, began to appear in newspapers across the country, and Hansen was soon appearing on television talk shows. Among his scientific peers, however, Hansen had lost a good deal of credibility. While many believed that global warming had indeed begun, few were willing to express such certainty about it. To make matters worse, in the media barrage that followed the Energy Committee hearings, Hansen was often misquoted or his words were taken out of context. Soon, as

Residents of Patiala, India, try to cope with the effects of a devastating flood that struck their town in 1988. Rising sea levels caused by global warming will flood many coastal areas around the world, disrupting the lives of millions of people.

Stephen Schneider writes in his book *Global Warming*, "people who had been toiling to resolve the uncertainty bit by bit without the world watching began to look for ways to shoot down Jim Hansen's now '99%' assertion."

AFTER THE ICE AGE

Although only a small minority of scientists do not believe in global warming, their dissenting opinions tend to get a disproportionate amount of play in the media. Thus, many atmospheric scientists are cautious about committing themselves publicly to the theory of global warming. Moreover, some of them are still smarting from a similar media circus that occurred 20 years ago, when many atmospheric scientists believed that the earth was headed toward a new ice age.

The Northern Hemisphere went through an unusually cold period in the early 1970s, leading some scientists to speculate that factory smoke and dust from farms was blocking sunlight and cooling the planet. This theory, along with fears for the ozone layer and for the planet's shrinking reserves of fossil fuel, helped ignite the environmental movement of the 1970s. During the early 1980s, however, there was a counterreaction. In an article in the December 1989 issue of *Forbes* magazine entitled "The Global Warming Panic: A Classic Case of Overreaction," writer Warren T. Brookes cited the earlier belief of many climatologists in a coming ice age and referred to atmospheric scientists such as Hansen as "calamitarians" who "always have something to worry us about."

Such criticisms not only ignore how much atmospheric science has advanced over the last 20 years but fail to mention

Atmospheric scientist James E. Hansen caused a stir within the scientific community when he testified before the Senate Energy Committee in 1988 that there was a 99% certainty that the earth was getting hotter.

that 2 of the 3 "calamities" predicted during the 1970s are still with us. Although oil seems to be flowing freely, the planet's reserves will probably run out within the next 100 years. Nevertheless, in the United States, funding for alternative energy research has dropped off drastically during the last decade. The limits placed during the 1970s on the use of chlorofluorocarbons—the chemicals that gobble up ozone in the upper atmosphere—never really cut down on global chlorofluorocarbon production, and by the early 1980s a dangerous hole in the ozone layer appeared above Antarctica.

Just as the public panic about ozone and dwindling energy reserves diminished during the 1980s, concern about global warming could easily evaporate by the turn of the century: A cool summer or two combined with a widely publicized scientific study refuting the warming predictions might convince people that the climate is "back to normal." The following chapters will show, however, that if industry continues to alter the atmosphere at the current rate, then the climate will never be truly normal again. It may take 20 years for scientists to prove irrefutably that the planet is warming, and it may take 100 years for global warming's most devastating side effects to manifest themselves. But if nothing is done to reduce the levels of greenhouse gases, future generations will pay for this generation's apathy.

Solving the problem of global warming demands foresight. As Crispin Tickell, British ambassador to the United Nations and author of the book *Climate Change and World Affairs,* is fond of pointing out, if a frog is put into a pot of boiling water, it will leap out immediately. But if a frog is put into a pot of cold water and the heat is turned up little by little, the frog will eventually boil to death.

Water vapor is an invisible gas. When the air cools, the vapor condenses into visible water droplets that form clouds. These clouds reflect both ultraviolet and infrared radiation.

chapter 2

A GLOBAL GREENHOUSE

Global warming has given the greenhouse effect a bad reputation, whereas in fact human beings owe their lives to it. Concerned by reports of impending doom, people have come to think of the greenhouse effect as unnatural, an artificial planetary plague brought on by industry and pollution. When Jean-Baptiste-Joseph Fourier, a mathematician who worked in Egypt under Napoléon, coined the term in 1827, he hardly meant for it to have such an ominous ring. Like the glass walls of a real greenhouse, Fourier argued, the earth's atmosphere insulates its inhabitants against the frozen outside. Human beings—and all other creatures on the planet—would wither and disappear without the planetary greenhouse, like orchids exposed to a northern winter.

If the earth were a bare sphere of stone hanging in space, it would absorb the sun's light and reemit the energy directly back into space in the form of infrared radiation. This process would give the planet's surface an average temperature of –2.2°F (–19°C). At that temperature, the earth would be as barren as Antarctica. But wrap a feathery blanket of atmosphere around that bare earth and the average surface temperature rises to 57.2°F

(14°C)—allowing the planet to sprout lush vegetation and the more complex life-forms that feed on those plants.

How does the atmosphere keep the earth warm? As one can tell when clouds move in front of the sun, the atmosphere does prevent a good deal of the sun's light from reaching the earth. About 25% is absorbed by the gases that make up the atmosphere, and about 25% is reflected back into space by clouds. In addition, 5% of the sun's energy strikes shiny, reflective areas of the globe, such as deserts and polar ice caps, and bounces back into space without ever being absorbed. The remaining 45% of the sun's light, however, does reach the surface of the earth. Visible light has very short wavelengths—short enough to slip past molecules of gas in the atmosphere, like skinny kids running through a crowd of adults. When sunlight is absorbed by the earth, however, it changes form. It rises as infrared radiation with wider wavelengths. The broader wavelengths of infrared radiation cannot squeeze past the gas molecules in the atmosphere and return to space. Long-wavelength radiation has a lower frequency, or energy level, whereas shorter-wavelength radiation has a higher frequency. It is this difference in frequency or energy levels between incoming solar radiation, in the form of visible and ultraviolet light, and outgoing heat, in the form of infrared radiation, that determines whether or not the radiation is blocked by the atmosphere.

A few gases—water vapor, methane, ozone, chlorofluorocarbons, nitrous oxide, and most important, carbon dioxide—are responsible for blocking the outflow of heat energy. These gases are the linchpins of the greenhouse effect; scientists call them greenhouse gases. Greenhouse gases have the crucial ability to absorb infrared radiation. When the gas molecules absorb infrared radiation, they gain kinetic energy, giving off radiation of their own and warming the air in the process. In this way, almost 88% of the infrared radiation or heat energy given off by the earth is absorbed by the lower atmosphere. Eventually, this energy escapes into space, but not before it has warmed the air along the way.

Although the sun is the original provider of heat, it is the atmosphere that traps that heat and raises the average temperature of the earth by a crucial 60°F (33°C). When people grumble because clouds cut off the sunlight, they forget about the invisible warmth that is reflected down in return.

THE EARTH'S CLIMATE SYSTEM

Everyone accepts the greenhouse effect: Global warming may be hotly debated, but the greenhouse effect is not. For more than 100 years, scientists have accepted it as a basic feature of the atmosphere, without which the earth would be uninhabitable. What scientists still do not fully understand, however, is how human beings affect the greenhouse effect.

It is easy to identify the basic elements of the greenhouse effect: the earth, a handful of gases, and the sun. But predicting how an increase in greenhouse gases might alter the greenhouse effect is as difficult as determining how a certain diet might affect a person's health. To do so, one has to understand how the entire planetary system functions.

When scientists analyze the earth's climate system, they must study the interactions of five basic elements: the atmosphere, the oceans, the cryosphere (the earth's ice caps, snowpacks, and glaciers), the biosphere (all plant and animal life), and the geosphere (the earth's crust). Any major change in one of these elements is liable to affect the rest. These interconnections are played out most swiftly and dramatically in the weather through hurricanes, typhoons, tornadoes, and thunderstorms. So many variables affect the earth's weather patterns—from the planet's orbit to ocean currents to volcanic eruptions—that anticipating their behavior is extremely difficult.

If, as countless studies have shown, industry is steadily filling the atmosphere with greenhouse gases, then one might expect global temperatures to rise proportionately with the increasing quantity of those gases. But the warming itself may be only the first step in a chain of atmospheric, geologic, and biological events—called a "feedback loop"—that could either enhance the original atmospheric warming or counteract it. Predicting what will happen depends on one's view of the earth's ability to regulate itself. As Stephen Schneider puts it: "Is our planetary life support system the delicately balanced, fine-tuned mechanism some fear or rather a resilient web of internal control systems that have and will continue to sustain life comfortably as they have for four billion years?" It is necessary to look at some of the basic processes or cycles that will determine the actual consequences of global warming. The interconnectedness of these processes makes predicting their effects very difficult.

It is not known, for example, how the amount of cloud cover will affect global warming. Water vapor is both the most important greenhouse gas in the atmosphere and the major obstacle to sunlight reaching the earth. On average, clouds cover half the earth, reflecting incoming sunlight back into space and bouncing infrared radiation back downward at the same time.

Clouds, therefore, both warm and cool the earth. If, as scientists predict, global warming continues to raise the temperature of the planet's ocean surfaces, as it has throughout the 1980s, then more water vapor should evaporate into the sky, creating more clouds. The final effect will depend on what kinds of clouds form from the new water vapor. If banks of low clouds build up, increasing amounts of sunlight will be reflected back into space, cooling the planet—possibly enough to negate the effects of global warming. If, on the other hand, banks of thin, high, cirrus clouds (with a strong greenhouse effect) begin to build up, they could speed the arrival of global warming disasters. Scientists still do not know which type of clouds to expect.

A major change in the earth's cryosphere would profoundly affect global climate. In the high reaches of the Northern Hemisphere, the Arctic ice sheet covers up to 6 million square miles (15 million square kilometers) of ocean, depending on the time of year. Unlike Antarctica, which is a stable, ice-clad continent, nearly half of the Arctic ice sheet melts every year. If global temperatures rise as predicted, more and more of the Arctic ice sheet will melt each spring and summer, and perhaps the ice sheet itself will disappear within the next half century. At present, the Arctic ice cools the earth's climate in two ways: by preventing the warmer water beneath it from coming in contact with the air and by reflecting 98% of the sunlight that falls on it back into space. As the Arctic ice sheet melts, therefore, both these cooling mechanisms will disappear. The ice sheet will be gradually replaced by dark sea water, which will absorb light rather than reflect it and warm the air on contact—further heightening global warming.

The biological activity of plants is an important factor in global warming. Plant activity is governed by two basic processes: photosynthesis and respiration. During daylight hours, plants take in CO_2, sunlight, and water and convert them into carbohydrates. At night, lacking the necessary sunlight, they respire the leftover CO_2 back into the air. Scientists have long assumed that as CO_2 increases in the atmosphere, the warmer temperatures would increase the global rate of photosynthesis. A flourishing of green plants would pull ever larger amounts of CO_2 out of the atmosphere, weakening the greenhouse effect. But, according to ecologist George Woodell of the Woods Hole Oceanographic Institute, a more ominous process may be at work at the same time: While monitoring a forest in Brookhaven, New York, Woodell and his colleague Richard Houghton found that as temperatures rose, the forest's rate of respiration grew at an astounding rate. Every time temperatures increased by 18°F (10°C), in fact, the trees respired CO_2 back into the air at twice their former rate, whereas the rate of photosynthesis grew only marginally. If this holds true, as the globe heats up, the biosphere could spew billions of tons of new CO_2 into the air every year, further warming the earth.

Even phytoplankton—the tiny single-celled plants that live in the upper layers of the ocean—can be important factors in a climatic chain reaction. Five years ago, a team of scientists headed by James Lovelock pointed out that certain types of phytoplankton produce dimethyl sulfide—a gas that the earth's atmosphere rapidly tranforms into sulfuric acid. Sulfuric acid particles in the air serve as the nuclei for billions of water droplets that form clouds. The more acid particles, the more water

A photograph of the surface of Mars taken by the unmanned Viking Lander 2 *in 1979. The photograph shows a thin layer of ice covering the rocky surface. Lacking sufficient greenhouse gases in its atmosphere, Mars is a very cold place, with an average temperature of –158° F (–120°C).*

droplets; the more water droplets, the more sunlight is reflected back into space, cooling the earth.

VENUS AND MARS

However confusing, atmospheric instability is a characteristic of a living planet. One need only look to earth's neighbors in the solar system to find planets that are very predictable and very dead—planets whose greenhouses have grown either disastrously hot or disastrously cold.

For years, Mars was considered the only other planet in the solar system capable of sustaining life. As late as the 1970s, in fact, many scientists believed that the deep furrows astronomers had noticed on the planet's surface were canals containing water. But when the National Aeronautics and Space Administration (NASA) was preparing to send the *Viking Lander* to the surface of

Mars in 1976 to settle the question once and for all, James Lovelock told them not to bother. The atmosphere of Mars is stable and unreactive, Lovelock told them; therefore, the planet must be dead. If it contained life, there would be some evidence of its unpredictable behavior in the atmosphere—some persistent chemical imbalance or illogical mixture of gases.

The NASA scientists went ahead with their plans, but they found no signs of life on Mars. Instead, they found a planet so cold (–158°F [–120°C] on average) that the poles are composed largely of frozen CO_2. Its atmosphere is too thin—100 times thinner than that on Earth—to retain the sun's heat. When the red planet's massive volcanoes were active in the distant past, they may have thickened the atmosphere enough with their smoke to create a greenhouse effect comparable to that on Earth—thick enough, perhaps, to warm the planet and sustain life. But Mars today is quiet, stable, and dead.

The sight of Venus also once inspired writers and astronomers to imagine life on its surface. With its swirling blue-green

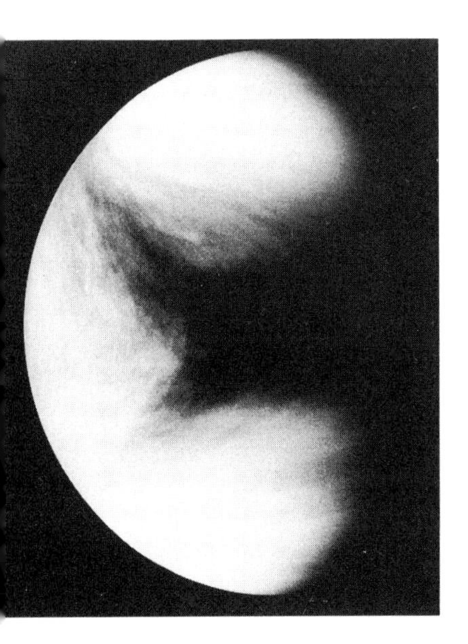

The swirling clouds of Venus, photographed by the Pioneer Venus Orbiter *in 1979. The Venusian atmosphere has such a high concentration of greenhouse gases, principally carbon dioxide, that its surface temperature reaches 840°F (450°C).*

atmosphere, Venus seemed to promise a benign, sensual climate. As late as 1918, Arrhenius himself suggested that "everything on Venus is dripping wet. . . . Only low forms of life are therefore represented, mostly no doubt belonging to the vegetable kingdom." But like the Sirens in the *Odyssey* who lured sailors to their island with music and then ripped them to pieces, Venus proved to have a nasty personality beneath its soft exterior. Insulated by an atmosphere 100 times thicker than Earth's, the surface of Venus reaches temperatures as hot as a kiln—around 840°F (450°C). Harsh clouds of sulfuric acid drift about in an atmosphere so dense that the pressure on the planet's surface is as heavy as at the bottom of Earth's deepest oceans. Such conditions may seem dramatic—a kind of planetary hell—but in reality they are static and lifeless: Scientists gave up imagining life on Venus 50 years ago.

Venus and Mars are similar to Earth in many respects. All 3 planets are approximately the same size and age (about 4.5 billion years old), they are made of the same material, and they orbit at comparable distances from the sun. Yet each has evolved differently. Whether Venus and Mars once supported life (in the case of Venus, probably not), their atmospheres have reached a kind of chemical equilibrium within which life cannot exist. Earth's quirky atmosphere, with its unstable elements, such as methane and oxygen, has managed to maintain a moderate greenhouse effect that has nurtured life for 4 billion years. But what if human beings are pushing their planet toward a new equilibrium? As the atmosphere loads up with increasing amounts of greenhouse gases, is Earth moving one step closer to Venus?

A tropical storm over the Indian Ocean, photographed by astronauts aboard the space shuttle Columbia *in 1990. The composition of the earth's atmosphere, and perhaps its climate as well, is slowly being changed by human activity.*

chapter 3

CARBON DIOXIDE: THE PRIME SUSPECT

If one thinks of the earth as an enormous living organism, then the breath of that great beast is carbon dioxide (CO_2). Astronomers estimate that when the planet was formed approximately 4.6 billion years ago, the sun was a quarter to a third cooler than it is now. On the earth as it is today, such a drop in solar-heat output would probably wipe out all life. Fortunately, the early atmosphere contained a far greater percentage of CO_2 than it does today. Spewing from the volcanoes and primordial fires that raged as the earth formed, these great clouds of CO_2 created an intense greenhouse effect that kept the planet warm enough to engender the first organisms. As the earliest algae eventually evolved into vast forests of complex plants, CO_2 was drawn out of the air through photosynthesis—leaving the world with a greener landscape and a milder greenhouse effect.

Today the earth's atmosphere contains only 353 parts per million of CO_2. But that amount has risen by 38 parts since the 1950s. To understand why these numbers are so crucial and why scientists are so concerned about the steady rise of CO_2 over this century, one has to understand CO_2's positive role in the greenhouse effect as well as its negative role in global warming.

When they were first formed, Venus, Mars, and Earth each had nearly identical amounts of carbon, but the three planets have stored their carbon in different ways. On Earth, carbon is the basic building block of life. As the binding element of every organic compound, it winds through the DNA in human chromosomes, the protein in skin, and the blades of grass in a park. This carbon, which seems so fixed and stable, is constantly being reused and reconfigured. Plants draw CO_2 into their leaves, transforming the carbon into fruits and more leaves; animals and human beings eat the leaves and fruits, reabsorbing the carbon into their body; eventually they die and decompose, returning some of that carbon to the ground and some back to the atmosphere.

This carbon cycle ties together the atmosphere, the biosphere, and the geosphere. The atmosphere contains only about two-fifths as much carbon as does the biosphere and about one-fiftieth as much as the oceans, but enough carbon passes through the atmosphere to make CO_2 the second most important greenhouse gas after water vapor. There is just enough CO_2 to keep people alive without frying them to a crisp. Venus and Mars, on the other hand, have kept their carbon under much stabler conditions. Nearly all of the carbon on Venus is suspended in the atmosphere—about 350,000 times more carbon per gallon of air than on Earth. This thick soup of CO_2 bakes the planet with intense heat. Mars, on the other hand, is too frigid for carbon to stay up in the air. Most of the red planet's CO_2 lies frozen as dry ice at the planet's poles—leaving a very weak greenhouse effect that makes even the sunniest spots colder than Antarctica.

If the earth's unique carbon cycle keeps people alive, it also leaves them extraordinarily vulnerable. CO_2, more than any other gas, is directly tied to the lives and actions of human beings. If the world's populations grow, if people continue to build factories, if they burn or cut down forests, if they drive rather than walk to work, they create CO_2. Every decision a person makes inevitably affects how much CO_2 is spewed into the atmosphere and how the greenhouse effect evolves. Consequently, CO_2 has become the primary focus of plans to curb global warming in the future. The ability to regulate how and when this gas is produced will determine whether it sustains or endangers future civilizations.

CO_2 was first discovered in the 17th century, by Flemish alchemist Johann Baptista van Helmont. For the next 200 years, scientists attempted to trace the sources of CO_2 and its concentration in various areas. Eventually, a rough outline of CO_2's origins emerged: Humans, animals, and plants breathed it out; plants breathed it in; burning wood, paper, oil, and coal released it into the air; and smokestacks belched it into the sky in great blasts.

Until the 20th century, CO_2 was mainly of interest to scientists. It existed only in trace quantities in the atmosphere—about 300 parts per million—and seemed harmless to human beings. It was not until Svante Arrhenius, at the turn of the century, finally connected CO_2 to the greenhouse effect that CO_2 suddenly took on larger significance, and even then to only a few visionary minds. If CO_2 is a greenhouse gas, Arrhenius reasoned in the April 1896 issue of *London, Edinburgh, and Dublin Philosophical Magazine*, then increasing the level of CO_2 in the air must bring about a "change in the transparency of the atmosphere." As the air blocked more of the infrared radiation rising from the earth, it would heat up, eventually melting great stretches of snow and ice near both poles. In his article, Arrhenius even speculated that as the polar ice cap shrank, exposing the ocean and tundra beneath, less light would be reflected back into space, allowing the earth to absorb more solar energy and to heat up even further.

POLLUTION PAST

Pollution is often considered an exclusively 20th-century problem—a result of too many cars, chemicals, and factories. But by the time Arrhenius made his predictions about the effect of rising CO_2 levels, half the damage had already been done. From the early 19th century on, the combined forces of the Industrial Revolution and European colonization were changing the face of the earth. By World War I, European nations had colonized more than 85% of the planet, cutting down forests to plant fields, digging mines to extract raw materials for home industries, and building cities to administer their new possessions. Carbon dioxide production worldwide increased suddenly and dramatically.

In recent years, the plight of tropical rainforests has gained worldwide attention. In the United States, people are shocked by the statistics: 33 acres of rainforest are chopped or burned down every minute, and up to 25% of the world's species will be extinct by the next century. But they forget that the country around them was similarly manhandled 100 years ago. Other than the Great Plains, which lay like a giant meadow from North Dakota through northern Texas, North America was essentially one continuous forest when European settlers first arrived.

A section of tropical rainforest in Guatemala is burned and cleared for farmland. This activity not only destroys the forest's biological diversity, it also releases massive amounts of carbon dioxide into the atmosphere, contributing to the global warming trend.

Trees were felled to make room for farms and to feed factory furnaces and the stoves and fireplaces of new homes. They were felled by the millions to make railroad ties and telegraph poles. Between 1800 and 1900, the United States lost most of its original woodland, a good deal of which has never been restored. For the nation, this was a great loss: first of the dense, old growth forests; then of those species, now extinct or nearly so, that the woods once harbored in great numbers—including wolves, passenger pigeons, grizzly bears, and red-cockaded woodpeckers. On a global scale, moreover, the deforestation of the 19th century—not only in North America but in China, India, South Africa, and Southeast Asia—dramatically disrupted the carbon cycle.

An old growth forest is a massive storage area for carbon, containing between 10 and 66 pounds of carbon per square yard (4 and 25 kilograms per square meter). Burning such a forest down, or using its wood to fuel factories, sends that accumulated carbon directly into the atmosphere. True, if another forest grows back in place of the original forest, the seedlings will eventually pull that carbon back out of the air. Not so with agriculture, however. A field of grain will pull down only about one-fifteenth of the original carbon stored at its site. By clearing the land to grow food, pioneers and colonists worldwide sent nearly 50 billion tons of carbon into the air during the 19th century. Although this massive discharge raised CO_2 levels by 20 parts per million between 1800 and 1900, it probably had little or no effect on global temperatures. But these early actions laid the foundations for the carbon pollution that would continue throughout this century, eventually raising the greenhouse effect to new intensities. Global warming, in some ways, is a late bill that humankind is paying for its expansionism and its hunger for cleared and cultivated land.

POLLUTION PRESENT

The environmental devastation of the 19th century was largely sparked by a population boom: The number of Europeans doubled between 1750 and 1850. In the 20th century, that population boom has continued around the world, consuming ever greater amounts of energy, spreading the destruction of forests to the Southern Hemisphere, and lofting carbon into the air at a rapidly increasing rate.

Between 1850 and 1950, the world's population doubled to 2.5 billion people. If people continue to procreate at this rate, there will be well over 10 billion people on the planet by the year 2050. Currently, 250,000 infants are born every day. In underdeveloped countries—where populations grow between 2% and 4% annually—these new souls have increasingly spilled into areas that were once wilderness. In Africa, the desperate need for fuel wood and land for new farms has destroyed more than 52% of all original tropical rainforests. Nigeria is expected to be completely deforested by the year 2000. In Asia, which has lost 42% of its original rainforests, the need for fuel wood and cropland, as well as the market for hardwoods, claims 620,000 hectares (a hectare equals a little less than 3 football fields in area) of Indonesian forest every year and has destroyed all of Bangladesh's primary rainforest.

When rainforest is cleared for pasture, it is most often burned; when it is gathered for fuel wood, its burning is delayed only for a short while. Altogether this clearing and burning of forest pumps more than 1 billion tons of carbon into the air every year. If Third World populations continue to grow and to remain underdeveloped, their appetite for wood will only escalate. According to the World Wildlife Fund, by the year 2000 more than half of the people in the Third World will not have enough fuel for their needs or will be consuming wood faster than they can grow it.

To avoid such a situation, the Third World is rapidly turning to First World technology and its primary source of energy: fossil fuels. In terms of pollution and carbon released into the atmosphere, however, fossil fuel is hardly an attractive alternative. Oil, coal, and gas provide 90% of the world's commercial energy, 70% of which is used by industrialized nations. People burn fossil fuels to generate electricity to run washing machines, stereo systems, blow-dryers, and electric toothbrushes. They burn fossil fuels to run lawn mowers, to propel jumbo jets, and to keep cities glowing as bright as a billion torches. Some experts predict that by the year 2000, more than 600 million vehicles will be moving over the world's roads, releasing CO_2 and other greenhouse gases with every blast of exhaust.

Every year, these "necessities" consume 3 billion metric tons of oil, 2.5 billion metric tons of coal, and 1.6 billion metric tons of gas (one metric ton equals 1,000 kilograms, which at the

The burning of fossil fuels at power plants such as this coal-fired unit in Eggborough, England, adds great amounts of carbon dioxide to the atmosphere.

surface of the earth weighs about 2,200 pounds). Over the last 30 years, all that burning has sent 80 billion metric tons of carbon into the air. Of the 8.5 billion metric tons of CO_2 that humans put into the atmosphere every year, 80% comes from the burning of fossil fuels. And the rate of pollution is accelerating. It has taken 90 years for CO_2 levels to rise from 300 to 350 parts per million in the atmosphere. But at current levels of pollution, it may take only 60 years for CO_2 levels to double to 700 parts per million.

TREE RINGS AND ICE BUBBLES

Since the Industrial Revolution, carbon has been pumped into the atmosphere at a fantastic rate. But what effect will rising CO_2 levels have on global temperatures? Will the CO_2 simply hang in the air and intensify the greenhouse effect, or will it be absorbed by the planet? Only about half the CO_2 emitted every year stays in the atmosphere. The rest sinks into the oceans or fertilizes the biosphere. But will the biosphere and oceans become saturated and actually begin emitting CO_2 themselves?

As usual, the best source for predicting the future is the past. Unfortunately, very few history books—much less scientific papers—were written 200,000, 10,000, or even 1,000 years ago. To assemble a picture of rising CO_2 levels in the 19th century, scientists have looked to trees: The type of carbon in tree rings directly indicates the level of CO_2 in the atmosphere. But for older, slower changes in atmospheric chemistry, science has had to turn to an even more unlikely source: ice bubbles.

Over the last 10 years, Russian, Swiss, French, and American scientists have drilled ice cores in the Alps, Greenland, and Antarctica that have proved to be atmospheric time capsules.

Up to a mile long, these cores reached down through tens of thousands of layers to ice nearly 500,000 years old. Within such ancient ice, perfectly preserved from the day that they were trapped by falling snow, lay millions of tiny air bubbles. By analyzing these bubbles, scientists have been able to determine the chemical content of the earth's atmosphere during successive stages of its evolution.

About five years ago, a team of French scientists began analyzing CO_2 levels in the air bubbles trapped in a particularly

Scientists at the Mauna Loa Observatory in Hawaii have been monitoring atmospheric concentrations of carbon dioxide for several decades. Carbon dioxide levels are steadily increasing, and it is believed that CO_2 is a principal cause of global warming.

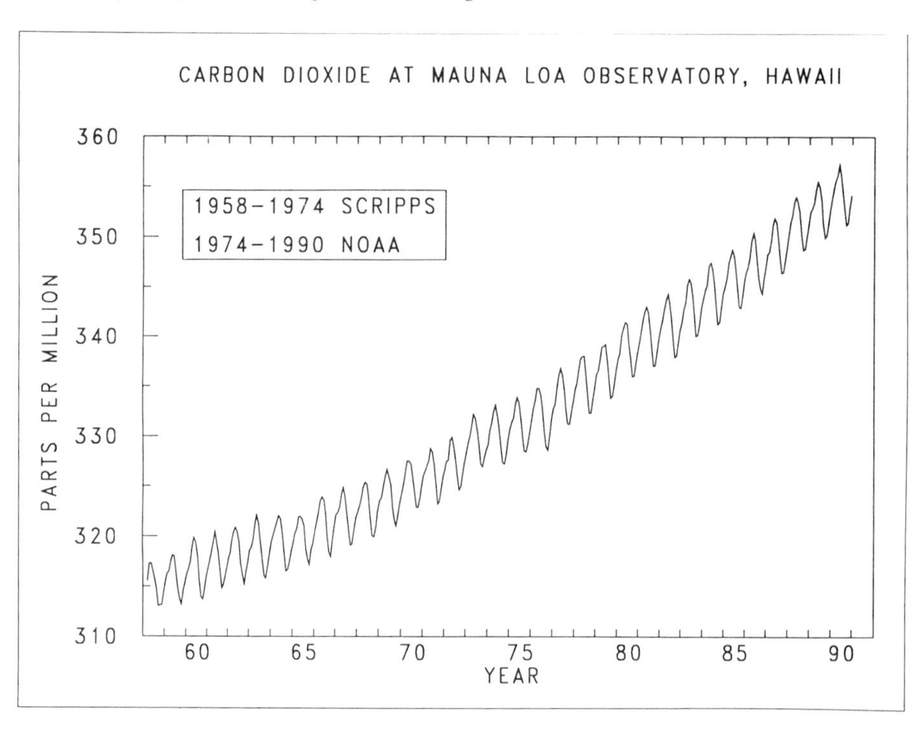

old ice sample collected by a Russian team working out of its Vostok base in East Antarctica. These scientists found that CO_2 levels fluctuated between 190 and 280 parts per million over the last several hundred thousand years. Seen as a graph on a piece of paper, this history of CO_2 looks like a meaningless zigzag, but plotted against a graph of temperature changes over that same period, it becomes clear that whenever CO_2 levels rise in the atmosphere, temperatures rise, and whenever CO_2 levels drop, temperatures drop. During an ice age, CO_2 levels hover at around 190 parts per million. During an "interglacial," as the warm periods between ice ages are called, CO_2 levels rise to 300 parts per million.

Some scientists argue that the Vostok chart actually offers some hope. If the world is indeed in the midst of an interglacial, as the charts suggest, then raising CO_2 levels artificially may slow down or soften the arrival of the next ice age. However, the next glacial period could take from 1,000 to 100,000 years to arrive. In the meantime, global warming could bedevil generation after generation—the handwriting, in the shape of a jagged double graph, is on the wall.

Human industrial activity adds many new chemical waste products to the atmosphere. Some of these chemicals are thousands of times more effective than water vapor or carbon dioxide in trapping the earth's heat.

chapter 4

THE GREENHOUSE GANG

Carbon dioxide is by far the most notorious greenhouse gas, yet if the greenhouse effect raises temperatures by 2° to 9°F (1° to 5°C) by the year 2050, it will not be due to rising CO_2 alone. A handful of other gases, whose sources range from termites to factories to hair spray, will have at least as much to do with it. As Stephen Schneider writes in *Global Warming,* "When all the human-produced trace greenhouse gases other than CO_2 are added up and their increases projected into the future, their collective greenhouse effect may add between 50 percent and 150 percent to the increase in greenhouse effect expected from CO_2 alone." Methane, nitrous oxide, chlorofluorocarbons, carbon monoxide, and ozone are the most important greenhouse gases after CO_2. However, the importance of these gases on the greenhouse effect is poorly documented and dimly understood. Even if human beings succeed in curbing CO_2 production in the next half century, limiting the effects of these trace gases will be difficult.

METHANE

When something smells bad in nature, it is probably releasing methane. Stagnant bogs, garbage dumps, belching cows—these are all sources of methane, or CH_4. (Every methane molecule contains one atom of carbon and four atoms of hydrogen.) Most often, when bacteria break down plant material—whether in the organic goo of a swamp or the belly of an animal—methane is released. Although the rotten scent associated with methane is familiar and at times hard to escape, it is much rarer than CO_2 in the atmosphere: Its concentration is only 1.7 parts per million. Today methane accounts for only 15% of the total greenhouse warming, but by the middle of the next century it could well be the most important greenhouse gas on the planet.

Since 1900, the concentration of methane in the atmosphere has nearly doubled, and it continues to rise at the rate of 1% every year—3 times the rate at which CO_2 is rising. More important, methane is an extremely effective greenhouse gas. Any given molecule can absorb 20 to 30 times as much infrared radiation as a molecule of CO_2. Unlike CO_2, a typical molecule of methane lasts only about 10 years in the atmosphere. As it floats upward, it combines with a hydroxyl ion (made up of one hydrogen atom and one oxygen atom) that uses the carbon atom in methane to form CO_2 and its hydrogen atoms to form water vapor. Both CO_2 and water vapor are not only effective greenhouse gases but at high altitudes water vapor freezes into clouds of ice crystals, which attract to their surfaces chlorine atoms from broken chlorofluorocarbon molecules, and chlorine is one of the ozone layer's most destructive agents.

Most of the methane released every year simply takes the place of methane that has decomposed—less than one-fifth of what is produced actually accumulates in the atmosphere. Because the products of methane decomposition are themselves greenhouse gases, however, methane increases the greenhouse effect both coming and going.

Most greenhouse gases come from sources that are generally thought of as pollutants: factories, burning forests, automobiles, etc. But a great deal of methane comes from more natural sources. Whereas deforestation and other burning releases 55 million metric tons of methane into the air every year, the world's growing herds of sheep, cows, goats, and wild herbivores send 80 million metric tons of the gas into the air. One cow can belch about 25 gallons (100 liters) of methane into the air every day. The vast rice paddies of Asia also release about 110 million metric tons of the gas annually; coal mining and leaks from

Both sheep and cattle release large amounts of methane, a greenhouse gas, as a waste product of digestion.

natural gas drilling and transportation activities release 35 and 45 million metric tons, respectively. Natural gas itself is 90% methane—58 million metric tons of it leak out of pipelines every year. And landfills do not pollute just the soil and groundwater; they pollute the air as well. On average, one-third of the contents of landfills consists of organic garbage, which releases 40 million metric tons of methane into the air every year. All told, human activities release more than a million metric tons of methane into the atmosphere every day. The United States leads the world in methane emissions, followed by India, China, the Soviet Union, and Canada.

Many atmospheric scientists, however, are less concerned about the methane emitted by human activities than about the methane emitted by the biosphere itself. Swamps and bogs already release 115 million metric tons of methane annually, and termites alone release 40 million metric tons. (For every person living on the planet today there are more than 300 pounds' worth of termites.) Because methane is released in proportion to the rate at which bacteria can eat and multiply, as the earth's soil grows warmer under an intensified greenhouse effect, the bacteria in the soil could grow more active and produce greater amounts of methane.

Such fumes will be nothing, however, compared to the quantities of gas that could come bubbling out of the planet's oceans, tundras, and inland seas under a warmer climate. Frozen beneath Arctic permafrost and hidden in the muddy bottoms of the Black and Caspian seas and the ocean floors surrounding the continents lie at least 10 trillion metric tons of methane. As oceans warm up and enormous tracts of frozen tundra suddenly turn wet and soggy under global warming, methane could begin

Bacteria in the digestive organs of these termites help the insects break down cellulose for food. Methane is released as a waste product. There are about 300 pounds of termites for every person living on the planet. Because termites prefer dead wood to living wood, the destruction of forests provides an ample source of food.

to escape into the atmosphere at a phenomenal rate. If such a runaway reaction occurs, pumping more than half a billion metric tons a year into the atmosphere, by conservative estimates, world temperatures will rise at least 5.4°F (3°C) more than already predicted by global warming experts.

Though dramatic, such scenarios are hardly farfetched. The ice cores drilled in Greenland and Antarctica contain records of similar events in the past. Atmospheric methane levels, like

CO_2 levels, rise whenever temperatures rise. In general, warmer climates put twice as much methane into the air as cooler climates, whereas CO_2 concentrations rise by only a third. Atmospheric methane levels doubled between the height of the Ice Age 18,000 years ago and the beginning of the 19th century and have doubled again since the Industrial Revolution.

Methane's natural sources make it extremely difficult to regulate. Although it is possible—but economically painful—to clean up factories and to increase the fuel efficiency of cars, people cannot eradicate waste, cork up livestock, or drain the world's rice paddies. Without rice—far and away the world's largest food crop—much of Asia would starve. And dozens of economies depend on the production of livestock. Fortunately, naturally produced methane emissions can be harnessed: In many Third World countries, composting animal dung supplies methane for stoves; in the United States, methane emissions from landfills have been successfully tapped. Moreover, when burned, methane emits half as much CO_2 as does coal to produce an equal amount of energy.

CHLOROFLUOROCARBONS

Developed in 1928 by Thomas Midgley, Jr., of the General Motors Research Laboratories, chlorofluorocarbons, known as CFCs, were originally designed to help cool the world down. At the time, GM needed a nontoxic, nonflammable gas to cool its refrigerators. Composed of atoms of chlorine, fluorine, and carbon, CFCs were the wonder chemicals of the age. Safe enough for Midgley to inhale them to blow out a candle, so stable that they take a century to break down in the atmosphere, CFCs immediately revolutionized the refrigerator and air-conditioning businesses. Between 1931 and 1945, worldwide production of CFCs soared from 545 to 20,000 tons. Today, whether people use their automobile air conditioners on blazingly hot days, congregate in climate-controlled art galleries, play indoor sports, or settle into airplanes bound for Cancún, CFCs keep their environment cool.

It is ironic, therefore, that CFCs are now considered one of the principal causes of global warming, second only to CO_2 in significance. Because CFCs are entirely man-made, their concentrations in the atmosphere are extremely low. Yet these few wisps of wonder chemical manage to block infrared radiation so effectively that they account for 17% of global warming. A single CFC-11 molecule can trap 17,500 times more heat than a CO_2 molecule; one molecule of CFC-12 can trap 20,000 times as much as a CO_2 molecule.

The only thing that can break CFCs down is ultraviolet (UV) light, but very little UV light reaches the lower atmosphere because it is blocked in the stratosphere by a thin layer of gas called ozone. A CFC molecule has to waft up to the stratosphere before rays of UV light will shatter it into its constituent elements: chlorine, fluorine, and carbon. Whereas CFCs are highly stable, chlorine and fluorine are extremely reactive, and they remain in the stratosphere and combine with other gases, such as ozone, the gas that protects living things from harmful UV light in the first place.

In the mid-1970s, a few chemists warned that this process could eventually destroy the ozone shield. In less than a decade, their fears were confirmed. In 1985, a group of British scientists announced that a hole in the ozone layer had been opening up

The rice paddies of Asia release more than 100 million tons of methane into the atmosphere every year.

over Antarctica every spring for the previous six years. By 1988, a committee of more than 100 experts announced that the ozone layer was quickly deteriorating across the globe.

Whether the danger is global warming or skin cancer, CFCs are bad news all around. The obvious solution, it seems, would be to ban production of these gases, but in the 60 years since they were developed, CFCs have insinuated themselves into every corner of the modern world. Through a series of industrial innovations—each more useful and convenient than the last—CFCs have been used in lightweight foams (such as Styrofoam), furniture cushions, automobile seats, and building insulation; as propellants in spray cans; and as solvents to clean microchips and electronic components. By 1990, CFC production had grown to more than 770,000 metric tons a year, 25% of which is used in aerosols, 19% in solvents, 16% in insulation and flexible foam, 12% in air conditioning, and 8% in refrigerants.

According to Stephen Schneider, "a phase-out of CFCs would be the single biggest step toward stabilizing the global atmosphere." Economically, however, a CFC ban will be a difficult pill to swallow. The United States currently leads the world in CFC production, followed by the Soviet Union, Japan, Germany, the United Kingdom, and Italy, and many other countries have substantial investments tied to them. According to the Alliance for Responsible CFC Policy, CFCs produced in the United States alone are worth $750 million annually; the services dependent on them are worth $28 billion, and equipment and products dependent on them are worth $135 billion.

Fortunately, the case against CFCs has been made strongly enough to override any economic concerns. After years of intense debate, representatives from 81 nations finally gathered in Helsinki in 1989 to declare that they would ban the production and use of CFCs by the year 2000. The accord was a welcome example of international cooperation, yet the CFC problem will still be with humankind for a while: Even after the year 2000 it will take the atmosphere 100 years—and countless ozone molecules—to cleanse itself of these wonder chemicals.

NITROUS OXIDE AND OZONE

Nitrous oxide (N_2O) is popularly known as laughing gas, but there is nothing funny about its presence in the atmosphere.

Coal fires, car exhaust, burning forests, and artificial fertilizers send about 6 million metric tons of N_2O (each molecule is composed of two atoms of nitrogen and one atom of oxygen) into the atmosphere every year, raising concentrations by 2% a decade. Since 1975, N_2O concentrations have risen from 295 parts per billion to 310 parts per billion. These may be tiny percentages, but N_2O is another potent greenhouse gas: Able to absorb 250 times more infrared radiation than can CO_2, an N_2O molecule can stay in the atmosphere for up to 200 years. Overall, N_2O accounts for 6% of global warming.

As with methane, N_2O emissions are difficult to control because they are partly tied to basic human needs. Nitrogen fertilizers are essential components of the modern farming methods that keep the world's booming populations fed. Since 1950, the use of nitrogen fertilizer has risen from 3 million tons to

Small amounts of nitrogen oxides are released from car exhausts. Although the atmospheric concentrations are small, nitrogen oxides are 250 times more effective at trapping heat than is carbon dioxide.

50 million tons annually. At first, this nitrogen is incorporated into plant tissue, but when the plant decays, the nitrogen combines with oxygen and rises into the air as N_2O. Methane and N_2O share one other worrisome trait: Under the right conditions, they can destroy ozone. It can take decades for N_2O to ride the air currents into the stratosphere where ozone dwells, but once there, it attacks the UV shield.

It may seem odd that a rise in ozone levels can contribute to global warming, but at the same time ozone is being depleted in the upper atmosphere, it is accumulating at a considerable rate in the lower atmosphere. When struck by UV light, chemical compounds released in car exhaust, paint fumes, spray deodorants, and other substances turn into ozone. The more UV light that makes it through the thinning stratospheric ozone layer, the more ozone is created in the lower atmosphere. Consequently, ozone levels in the troposphere have doubled over the last century. While ground-level ozone may prevent a few rays of UV light from striking your skin, it is more trouble than it is worth. Ozone causes sore throats, burns the eyes, and poisons the lungs.

An official of the Department of Agriculture examines withered stalks of corn, caused by drought. Global warming and the resulting changes in weather patterns may bring such conditions to much of America's most productive farmland.

chapter 5

THE IMPACT OF GLOBAL WARMING

Civilization has been constructed around a more or less predictable climate. Generations of people have built fishing villages, ports, and cities on the coasts, assuming that water would ebb and flow to the same shorelines every year. They have built dams to catch rivers and streams that flood with snowmelt every spring, dikes to withstand raging seas, and irrigation systems to carry water to areas that never see a cloud. They have planted crops according to seasonal rainfall and frost and colonized islands, thinking them pristine and safe from outside interference. In the 20th century, however, these certainties have been called into question. Global warming will likely cost human society dearly.

Over the last 10,000 years, temperatures have remained remarkably stable across the globe—changing by little more than 2°F (1°F equals approximately 1.8°C) on average. Even during the Little Ice Age, which lasted from the 14th to the 19th century and left a legacy of advancing glaciers, stunted crops, and death from exposure in Europe, mean temperatures were little more than 2°F colder than they are today. Global warming could change average temperatures five times as much as the Little Ice Age did—though in the opposite direction. Over the next century, the rate of global warming should follow a steep upward curve—heating up just a bit in the 1990s and progressively more each decade thereafter—until the earth's temperature rises to the sweltering levels that occurred hundreds of thousands of years ago. Such catastrophic temperature changes do occur every 80,000 or 90,000 years, when the earth's orbit swings it far enough away from the sun to create an ice age, but the biosphere has thousands of years to prepare for it. Plants gradually evolved hardy genotypes as ice sheets advanced, animals migrated slowly south from century to century, and early nomadic hunters passed on new methods of building shelter. How will the planet and human society adapt with only 50 years to prepare for an equally dramatic change?

Although most scientists agree that the globe will heat up between 2° and 9°F by the year 2050, the computer climate models that generate these warming predictions are still too clumsy and simplistic to predict regional climates. A typical model can predict climate change across an area of hundreds of thousands of square miles. All the countries of central Europe might be subject to the same climate prediction even though each country's seasonal temperature, rainfall, and snowfall patterns are unique. Using even the best climate model to make such specific forecasts is a hopeless task at present.

Computer climate models can, however, predict climate trends across the globe under higher temperatures. They can determine that global warming will make summers drier in midlatitude areas across the planet—even though they might not be able to foretell the average summer rainfall in California in the year 2050. These predictions sketch a general picture of how global warming will affect climates across the face of the globe and how such changes might disrupt civilization and the biosphere.

The earth's northern latitudes should feel the most dramatic effects of global warming. As the Arctic and Antarctic ice packs shrink and the exposed tundra and seas absorb heat that was once reflected back into space, the poles will probably heat up at an accelerated rate. Climate models predict that if global warming raises temperatures by 3.5°F on average, then temperatures in the Arctic could warm up by as much as 7°F. (Whereas Arctic winters will be especially warm, Arctic summers will grow only slightly warmer.) The equatorial regions, on the other hand, will be less subject to global warming, perhaps

The forward edge of a glacier, where large chunks of ice calve off and float out to sea under the influence of warmer temperatures. Much of the world's supply of fresh water is locked up in polar ice; global warming will melt this ice and raise the levels of the oceans.

heating up a degree less than the global average. As the world's oceans are fed by melting glaciers and ice caps, sea levels will rise by as much as 3 feet (1 meter) within the next 100 years.

As temperatures rise, the oceans will evaporate more quickly, creating more clouds and more rainfall. According to Jonathan Wiener, author of *The Next One Hundred Years,* 500,000 cubic kilometers of water rise and fall between the oceans and sky every year—an amount that could increase by 25,000 cubic kilometers under global warming. (One cubic kilometer is equal to a volume of space about 12 city blocks long, 12 blocks wide, and 12 blocks high.) However, this extra rain may fall when and where people least need it. Subtropical areas, such as India, may be hit with heavy floods and monsoons, and vulnerable farmlands in temperate climates will experience droughts.

At high latitudes, in areas such as Canada, warmer temperatures will mean longer growing seasons and more rainfall. Soviet climatologist Mikhail Budyko, one of the earliest investigators of global warming, believes that this process will bring food and prosperity to vast areas of the globe that have been desolate for thousands of years. Central Asia, he predicts, will receive 50% more rain, and the Sahara may receive 12 inches (30 centimeters) more rain every year by the late 1990s. Altogether, he believes that increased rainfall and agricultural productivity (rising CO_2 levels will also increase the rate of photosynthesis) should raise world food production by 50% in the year 2050. Such optimism is rare among scientists studying global warming.

Whereas mountain snowpacks in midlatitude areas such as California and Colorado should melt earlier—meaning more rainfall and river water in the spring—midsummer, when crops need the water most, will be drier. Together with the increased probability of heat waves, these conditions could lead to frequent droughts and spreading deserts in areas such as the Great Plains and Western Europe.

Given these general outlines, it is possible to imagine endless environmental disasters caused by global warming. Will Antarctica's western ice shelf drop into the ocean? Will California become a true desert? The initial effects of global warming need to be looked at in more detail.

THE POLES

The first journalists to write about global warming inevitably focused on Antarctica. Extending from the western edge of the continent, an ice ledge a third of a mile thick lies over the ocean, supported and restrained from floating away by a few islands. This ledge, in turn, keeps the rest of the Antarctic ice sheet behind it from surging into the ocean. As global temperatures rise by up to 9°F, this shelf could conceivably break apart, allowing enormous chunks of ice to flow into the ocean. In the space of a century, these chunks of ice could displace enough water to raise global sea levels by more than 15 feet (approximately 3 meters), flooding Boston, submerging Miami, and engulfing Bangladesh and a good deal of Southeast Asia. Most glaciologists now believe that Antarctica's western ledge is fairly stable—at least enough to last a few more centuries. Some climate models even suggest that rising temperatures will cause new snow to fall on Antarctica, enlarging it and making sea levels drop for a time. But the possibility for catastrophe exists.

Although it has received less public attention, the Arctic ice cap now worries scientists more than Antarctica. More than half of the Arctic ice pack melts every summer, growing to full size again in the winter. Melting the ice cap entirely, as some scientists fear global warming will do within the next 60 years, would have very little effect on sea levels, but it could disturb the global climate and oceanic currents. Arctic ice not only reflects a great deal of sunlight back into the atmosphere but also absorbs a great deal of heat as it melts every year. Weather in the Northern Hemisphere would be drastically altered without this seasonal interchange of heat and cold. Furthermore, the sinking of heavy salt water, which Arctic ice expels every year as it freezes, drives nutrient-rich water up from the depths of other oceans. These nutrients are consumed by hungry plankton all over the world. Without a yearly freeze to drive it, this nutrient cycle might stall, starving a good deal of the world's marine life.

Residents of Dhaka, the capital city of Bangladesh, rescue a vehicle from local floods. If global warming proceeds far enough, warmer temperatures will melt polar ice, raising sea levels and flooding many low-lying areas of the world.

RISING WATERS

Approximately one-third of the world's population lives within 100 miles (160 kilometers) of a coast. If global sea levels rise only 2 feet (.6 meters) over the next century—a conservative estimate—the cost to human society will be enormous. Bangladesh, a country that is nearly at sea level, would lose millions of acres to flooding. If sea levels rise by 6 feet (2 meters), 28% of the country would be flooded, and 27% of the population would be displaced. The Maldive Islands, southeast of India, with a population of 214,000, would fare even worse. At an average of only 6 feet above sea level, the islands could be almost completely submerged if the worst-case scenarios for global warming come true. Without seawalls, many of Indonesia's seaside rice paddies would be destroyed by salt water, and groundwater reserves for many coastal cities around the world would become contaminated with salt, rendering the water undrinkable. In the United States, 20% to 45% of all wetlands would be lost to a 2-foot rise in sea level, and between 2,000 and 6,000 square miles (5,200 and 15,000 square kilometers) of dry land would succumb to flooding.

STORMY WEATHER

Like the frog slowly coming to boil in the pot, the average person may hardly notice global warming's impact—after all, it will take a full lifetime to play out. But as the atmosphere heats up, it should provide people with a few sharp jolts in the form of more frequent hurricanes, heat waves, cyclones, and typhoons.

Heat and moisture are the fuels that feed atmospheric turbulence. A typical typhoon or hurricane develops only over

A warmer climate will probably produce more violent storms around the world. This possibility, coupled with higher sea levels, could devastate coastal areas and the dense human populations that live there.

warm tropical seas. As the world's oceans heat up over the next few decades, tropical storms may become increasingly common. They may also grow increasingly violent. Most storms today extinguish themselves at the height of their fury by coming into contact with cold ocean or land masses. In warm, partially protected oceans such as the Gulf of Mexico, hurricanes could grow up to 60% more intense as the upper layers of the ocean grow warmer. Higher surface sea temperatures could also extend the range of tropical cyclones as far as 60 degrees south in areas such as Australia. Coupled with rising seas, such storms could bring further disasters to the world's coastal cities. Tsunamis and torrential floods that happened only once every 50 years may now happen every 20 years.

THE WATER SUPPLY

With 500,000 cubic kilometers of water falling on the earth every year, and that number rising with global warming, one might think that there will be no shortage of water over the next 50 years. That depends on where a person lives. California is a good example: Although it is thought of as a lush paradise, the southern half of the state is largely semiarid cactus country. If Los Angeles County can support a city at all, it is because water is pumped to it every day from the Sierra Nevada by way of a 337-mile-long (544-kilometer-long) aqueduct. To the north, the vineyards and orange plantations of the Central Valley are well fed by rivers. Both these sources, however, are extremely vulnerable to global warming. During the brief heat waves of 1988, the fraction of California's electricity provided by hydropower dropped from 20% to 7%. Los Angeles had to resort to water rationing. Under global warming, the frequency of such heat waves will rise, just as the number of disastrous floods will rise, and water rationing will become common. Although it may rain more heavily in the springtime, melting the Sierra's snowpacks and filling the state's reservoirs early, there may not be enough water left in the dog days of midsummer, when crops need it most.

Similar scenarios will be repeated in vulnerable areas around the world, in countries less equipped to deal with water shortages. Of all the fresh water taken from streams, 75% is used for irrigation. As droughts hit, reservoirs used for irrigation will inevitably be diverted to more vital uses. To save their crops, farmers will have to find ways to pump more groundwater—if they can afford it.

A final threat to the water supply and to mountain communities comes from glaciers. As temperatures rise, glaciers will melt more quickly, throwing hydroelectric systems out of whack and increasing the likelihood of glacial floods, which occur when melted water stored beneath the ice suddenly breaks forth in a torrent. Whereas dams and mountain cities in developed countries are prepared for such disasters, in Third World countries some dams could crumble, and their power stations could fail.

AGRICULTURE

It is easy to see why economists from Canada and the Soviet Union are less concerned about global warming than their counterparts to the south. If the north gets the bulk of the heat and precipitation, will not the breadbaskets of the world simply shift north (and, conversely, south in the Southern Hemisphere)? Will not Nebraska's corn belt simply move to Ontario, and China's fields of grain to Siberia? Moreover, some agronomists argue that rising carbon dioxide levels will increase the growth rate of all crops; new agricultural technology will increase crop yields by as much as 50% by the year 2050.

Perhaps, but a number of negative side effects of global warming promise to cloud this rosy vision and to decrease worldwide food supplies. While it is true that grain belts should move north as temperatures rise, northern soils are much poorer than those to the south. In North America, the Environmental

Protection Agency (EPA) estimates that corn and soybean crops in Minnesota and farther north will increase, whereas crop acreage in the Great Plains will decrease by 4% to 22%, reducing dryland yields of corn, wheat, and soybeans in this region by as much as 80%. At the same time, although CO_2 may help increase crop growth rates, it will also increase the growth rates of weeds and other competition. Improved agricultural technology will be necessary to compensate for the phenomenal rate at which the world's topsoil is being used up—more than 16 tons per hectare every year—and for the increasing resistance of crop-devouring insects to pesticides.

The rising heat itself could be the most serious threat to world agriculture. Frequent heat waves will not only start more forest fires and dry up more streams but will also ruin entire corn harvests. Under similar circumstances, during the years of the Dust Bowl in the late 1920s and early 1930s, the United States produced 50% less wheat and corn. In 1988, the heat was strong enough to destroy 40% of the nation's corn. Such problems will be felt much more keenly in the Third World. According to a study by Anne and Paul Ehrlich cited by Worldwatch Institute, a more erratic and drought-prone climate could cause 2 serious depletions of grain stocks each decade, resulting in the loss of 50 to 400 million lives.

HABITAT DESTRUCTION

People can always move north; sometimes they can even take their crops with them. But forests, wetlands, and coral reefs have to sit and take the heat. Stephen Schneider points out that as the last Ice Age ended, forests also migrated northward, but they had thousands of years to do it. If Mississippi's climate moves to Michigan in the space of 50 years, how will the forests survive? Beech, birch, and maple trees, and a few other important species in the Southeast will almost certainly die off. To keep the country's forests from becoming wastelands, the EPA estimates that U.S. reforestation efforts will have to be increased two- or threefold. The high latitudes will hardly escape these troubles. A study conducted in Ontario has shown that the recent rises in temperature have already increased forest fires and droughts, dried out soils, and stunted forests.

Whereas forests can be replanted with great effort, entire ecosystems—with their indigenous animals and small

Coral reefs flourish in the warmer surface waters of the world's oceans. Hotter seas caused by global warming could push the reefs beyond their tolerance for heat and destroy them. This would be catastrophic for the many species of fish that feed there.

plants—cannot. Although the global rate of species extinction is already higher than it was when the dinosaurs were killed off, it promises to climb much higher over the next few decades. The loss of wetlands will certainly reduce many fish, shellfish, and bird populations, and many species of marine and lake fish will not be able to tolerate warmer waters. Animals living on islands or in national parks hemmed in by civilization will either be forced to adapt or die off.

Coral reefs, where the majority of the world's ocean species live, may prove very sensitive to heat. Reefs grow close to the surface of the ocean, where waters are warmest; they live near the upper limits of their tolerance to heat. A recent study by marine researcher Ernest H. Williams of the University of Puerto Rico showed that exceptionally warm seas have already taken their toll on coral reefs around the world. In reports collected from scientists working as far away as Japan, Hawaii, Kenya, and the Maldives, Williams found widespread evidence of coral bleaching—a response to environmental stress that can eventually lead to the death of a reef. If this trend continues, many of the world's coral reefs will not last into the next century.

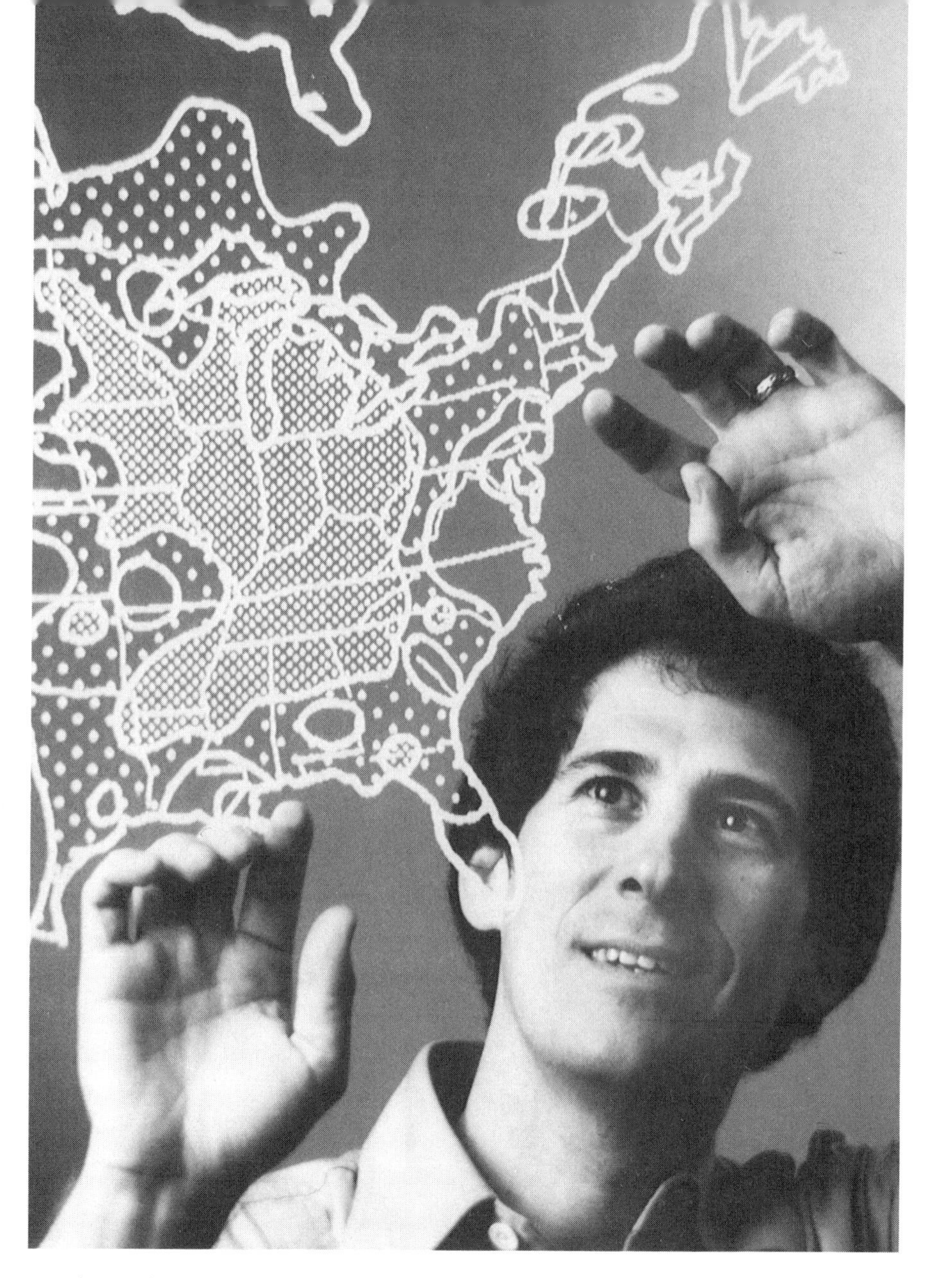

Stephen Schneider of the National Center for Atmospheric Research is one of a number of scientists trying to educate the public about the problem of global warming.

chapter 6

CLIMATE PREDICTION

Climate prediction is both a very old and a very young science. Every farmer in the United States is familiar with *Poor Richard's Almanac,* which forecasts seasonal rainfall, snowfall, and temperature. Though such predictions have some reliability, the daily weather in any given place is still hard to forecast. As meteorologists often put it, a butterfly beating its wings in Paris could change the path of a storm in Dallas. But climate—the average weather in a given area over the course of many years—is fairly dependable.

By studying written records, glacial remains, fossil plants and animals, and ice cores, scientists and historians have gradually been able to reconstruct a picture of the earth's earlier climate. The earth was warmer when some of the dinosaurs lived 100 million years ago and much colder 18,000 years ago. Broad trends have emerged. Over the course of 100,000 years, the earth's orbit around the sun changes shape. Over the course of 41,000 years, its axis tilts between 22.5 and 24 degrees away from the plane of its orbit. Over the course of 24,000 years, a third cycle, called precession, further affects the planet's position.

Periodically, these three cycles coincide to cool the Northern Hemisphere enough to start an ice age.

Predicting climate trends—and the human impact upon them—in the near future, however, is still a tricky proposition. The most obvious solution is to do what generations of farmers have done; that is, to learn from the past. Modern climatologists call this the paleo-analogue method.

Bubbles in Antarctic ice cores have shown that as levels of greenhouse gases rose in the past, so did global temperatures. Climatologists use these figures to predict how global temperatures will rise over the next few decades at current rates of greenhouse gas emission. To determine how this warming will affect global climate, they examine past climates when temperatures were equally high. During the Pliocene epoch, for example, temperatures were 5° to 8°F higher than they are today. Fossil evidence of that period reveals levels of rainfall and types of vegetation that can help scientists predict how regional climates will change if temperatures rise by 5° to 8°F in the next 100 years.

Although helpful, paleo-analogues can give scientists only a rough, often unreliable idea of global climate change. Critics of global warming theory argue that the parallel between rising carbon dioxide levels and rising temperatures is misleading. Antarctic ice bubbles show that when CO_2 levels rose by 50%, temperatures rose by 20°F. If, as the records show, CO_2 has risen by 25% since the 19th century, then global temperatures should have risen by 10°F. Yet temperatures have risen by only .5° to .9°F. In the past, climate changes were part of natural, long-term heating and cooling trends—hardly perfect analogues for an abrupt, man-made warming. By the same token, the evidence of past climate changes is too limited, and the factors that determine climate are too varied, to assume that the past will simply repeat itself.

MODELING THE EARTH

Dissatisfied with what the past can tell them, scientists have designed general circulation models (GCMs). GCMs are computer programs based on mathematical equations that try to duplicate the behavior of large-scale meteorological processes. Programmed into supercomputers, GCMs generate imaginary

Researchers at the Goddard Institute for Space Studies, a division of the National Aeronautics and Space Administration, study the predictions of a global circulation model, or GCM, a computer program designed to forecast climate change.

spherical models of the earth, inscribed with rough outlines of the world's oceans and land masses, with the entire surface subdivided by a grid. Each box in the grid has sides hundreds of miles wide and more layers of boxes stacked above it, reaching into an imaginary stratosphere. The atmosphere, oceans, and continents of this idealized planet interact according to simple physical laws that dictate the behavior of solids, fluids, and gases on a rotating sphere. Water vapor rises, condenses, and falls back down as rain. The globe heats up from mathematically simulated sunlight, driving winds that in turn push the ocean's surface currents. If scientists program cooler temperatures into the GCM, ice spreads forth from the poles, reflecting more sunlight and automatically cooling the earth even more. Such complex formulas strain the capabilities of even the biggest and fastest computers. Simulating 1 day's worth of weather on a GCM can take half an hour; 1 year's worth requires more than 500 billion calculations. The best way to test a GCM is to have it predict present climate. Given today's global atmospheric temperatures, for example, how warm should the surface of the oceans be? The four most powerful GCMs—located in Colorado, New Jersey, New York, and Great Britain—can answer such questions accurately. They can reliably reproduce today's jet streams, rainfall patterns, and seasonal temperature changes. Given a few basic statistics, they can even simulate the climate 18,000 years ago—consistent with the story that fossil plants, animals, and ancient air samples have already told.

Predicting the future with a GCM can be complicated. In Colorado, New Jersey, New York, and Great Britain, the climatologists tell their GCMs to predict global temperatures if CO_2 levels double, and they all answer nearly the same way: Temperatures will rise 2° to 9°F. Those same models, however, predict that, given the amount of CO_2 already in the atmosphere, global temperatures should have risen 2°F this century. Once again, the actual rise in temperatures this century—.5 to .9°F—confounds the predictions. What could the models be doing wrong? And do their inaccuracies mean that global warming will be easier or harder to live with?

This computer image shows how global circulation models first impose a grid on the surface of the earth to break down patterns of atmospheric circulation into manageable blocks of data.

CLOUDS

Clouds are a very unpredictable variable within the GCM. They come in so many shapes and sizes, at so many altitudes, and are so vulnerable to winds that there is no way for a climate model to predict their behavior accurately. Yet the cumulative impact of clouds on global temperatures could be up to 100 times that of greenhouse gases, and their influence is sure to grow as ocean surfaces begin to evaporate up to 3 times as quickly under global warming. High clouds increase the greenhouse effect; low clouds decrease it. Knowing what kinds of clouds form as temperatures rise is crucial to determining the extent of global warming.

Climatologists do have a makeshift way of programming clouds into their GCMs. To each box in the model's grid they assign parameters defining the average cloudiness of that box according to its average temperature and humidity. The initial prediction of relative cloudiness works well, but the models have a hard time assessing the further effects of the clouds themselves. First of all, the models themselves disagree. When cloud activity is taken into account, the predictions for global warming vary by a factor of two. Richard Lindzen, of the Massachusetts Institute of Technology (MIT), has been one of the most outspoken critics of the models' treatment of clouds. He has argued that rising temperatures could actually dry the upper atmosphere. Particularly in the tropical latitudes, water vapor may rise quickly into the upper atmosphere through a process called deep convection. Once there, it will rapidly condense and precipitate more quickly, drying the upper atmosphere and thinning high cloud cover. Lindzen believes that this process could decrease estimates of global warming sixfold.

A 1988 study of satellite data on cloud cover appeared to confirm such criticisms of climate models, showing that clouds have a net cooling effect on the earth's atmosphere. In a recent issue of the scientific journal *Nature*, however, the head of the satellite study, the University of Chicago's Veerabhadran Ramanathan, revealed dramatic new evidence that contradicted his earlier conclusion. In his study, Ramanathan found that the extremely humid atmosphere above the tropics traps twice as much infrared radiation as does the dry atmosphere above the poles. Ramanathan attributed this increased greenhouse effect directly to water vapor, implying that the increased cloud cover was a definite warming factor. Ramanathan did find some evidence of the atmospheric drying and cooling predicted by Lindzen, but it was minor compared to the heating effect.

OCEANS

The world's oceans, which take up two-thirds of the planet's surface area, are the biggest unknowns in any discussion of long-term climate. Clouds are unpredictable, but at least they form and circulate on a time scale that one can measure easily. One can watch the progression of a tropical storm day by day and issue reasonably accurate warnings to those living in its path. But it takes many lifetimes to change the circulation patterns or the average temperature of an ocean. Two and a half miles (four kilometers) deep on average, the world's oceans are enormous heat sinks. It takes an ocean several hundred years to warm up a few degrees, even under the worst-case global warming scenarios. Climate modelers, such as James Hansen, believe that for global warming to occur the atmosphere has to warm only the upper 300 feet (100 meters) or so of the ocean, but the stubborn coldness of the oceans could delay the full effects of global warming by 10 to 100 years.

The oceans are also powerful CO_2 sinks: They absorb about 3 billion tons of CO_2 every year. Because ocean waters take anywhere from 100 to 1,000 years to mix, the surface waters absorbing all this CO_2 may be nearly saturated by now. Some scientists suggest that if atmospheric CO_2 levels continue to rise, the oceans will absorb progressively less of it, raising the level of atmospheric CO_2 further still.

Ocean currents are still poorly understood, but they can have dramatic effects on regional climates. As Andrew Solow, a critic of global warming theorists, points out, atmospheric conditions change rapidly across thousands of miles at a time, where-

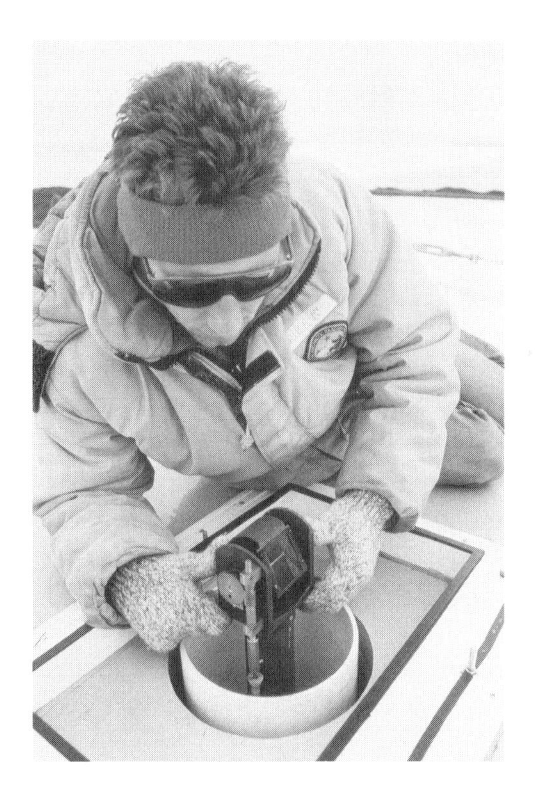

A scientist working in Antarctica prepares a device that will monitor the ozone layer.

as oceanic conditions change slowly across tens of miles. Areas where cold, deep waters and warm surface waters hardly mix may allow those surface waters to heat up very quickly under global warming; areas where cold, deep ocean currents rise to the surface may warm much more slowly. Such regional differences in temperature will then drive wind patterns that will determine rainfall and other aspects of climate in those areas. At the same time, if global warming causes certain ocean currents to halt altogether, the results could be disastrous. In Europe, around 11,000 years ago, during what geologists call the Younger Dryas, a temporary shifting of the Gulf Stream may have brought on a mini–ice age, lowering average temperatures by as much as 9°F.

All these interrelated mysteries limit the ability of GCMs to accurately predict future climate. While there are computer models for oceans, complete with multiple temperature layers and ocean currents, coupling them with atmospheric models has been frustrating. As climate modeler Syukuro Manabe told *Time* magazine in 1987, "Even though the atmospheric model and ocean model work individually, when you put them together, you get crazy things happening. It's taken us 20 years to get them together, and we're still struggling." In the years since, some of the bugs have been removed, but oceans are still an unsettling question mark in any climate prediction.

Another problem for climate modelers concerns the quality of the data fed into the computer. Most of the thermometers used to monitor temperatures worldwide are located in cities, which are significantly warmer than surrounding rural areas. In 1988, in the midst of the media storm surrounding global warming, the problem of "urban heat islands" was often brought up to refute the record of rising temperatures. However, once conservative corrections were made to compensate for city heat, the record still showed temperatures rising by .9°F this century.

Critics have also complained about many inconsistencies between the historical record and the theory of how global warming is supposed to work. As *Forbes* magazine, in its cover story "The Global Warming Panic: A Classic Case of Overreaction," pointed out, "Most of the past century's warming trend took place by 1938, well before the rise in CO_2 concentrations." Again, the heat-absorbing ability of the oceans might account for this delayed reaction. Andrew Solow has also pointed out that the geographic distribution of global warming so far has not been consistent with climate model predictions. The Southern Hemisphere, tropics, and the oceans along the equator have warmed up more than the Northern Hemisphere—contrary to expectations.

The world's most prestigious scientific journals have been filled with conflicting studies about global warming. In the fall of 1989, MIT's *Technology Review* released a study showing that worldwide ocean temperatures had risen only 1° to 8°F since 1860, concluding that "there appears to have been little or no global warming over the past century." In March 1990, two conflicting studies appeared: One, in *Nature*, declared that natural variations in climate are "insufficient to explain the observed global warming during the 20th century." The other study, in *Science*, reported that satellite monitoring of the earth's lower atmosphere during the 1980s showed no evidence of warming. However, one of its coauthors, Roy Spencer, later told the *Boston Globe* that a decade of data is "quite meaningless" in terms of long-term climate trends. Most recently, two British climate researchers, who have compiled a record of temperatures since the middle of the last century, declared that the 1980s were clearly the warmest decade on record and that 1990 was the hottest year in more than a century. Spencer has since said that his satellite records for the years from 1979 to 1988 correlate well with the British climate study.

Taken altogether, this barrage of contradictory studies seems to tarnish the credibility of GCMs. As Christopher Flavina and Nicholas Lenssen have pointed out in *Worldwatch* magazine, however, "This apparent controversy is largely a media creation, resulting from disproportionate press coverage of a small number of dissenting scientists. Beneath all the hoopla, the basic consensus about global warming has strengthened as pieces of a very complex scientific puzzle are fitted together." The most convincing evidence of this consensus was the publication in December 1990 of *Climate Change,* the official report of the World Meteorological Organization and the Intergovernmental Panel on Climate Change (IPCC). Composed of several hundred working scientists representing 25 countries, the IPCC spent 2 years creating a definitive description and prediction of the global warming trend. In the book's executive summary, the editors declared themselves "certain of the following": first, that "there is a natural greenhouse effect," and second, that "emissions from human activities . . . will enhance the greenhouse effect, resulting on average in an additional warming on the Earth's surface."

The contributors to the IPCC volume agree that many of their predictions are shaky or vague, that the GCMs are flawed, and that there is a desperate need for further research into oceans, clouds, and other variables. But they also stress the fact that there is no time to wait for better crystal balls.

Too many people, too many automobiles, too many and too much of everything. Creating the energy necessary to sustain the modern urban life-style may alter the world's climate—is this really the way anyone wants to live anyway?

chapter 7

SHADES OF WORLD GOVERNMENT

According to MIT economist Henry Jacobi, a problem must have four characteristics to merit a swift political response: It must be *serious*, *certain*, and *solvable*, and its effects must be felt *soon*. Global warming, unfortunately, fails to meet at least three of these four criteria. The effect that rising greenhouse gases will have on the global climate is anything but certain, and according to scientists such as Richard Lindzen, it may not even be serious. Most scientists agree that although rising temperatures may already be damaging coral reefs and a few other isolated areas of the environment, the full effects of global warming will not be felt soon, and they may take a century or more to appear. As for the fourth criterion, even if the world's governments take immediate action, the problem of global warming is not truly solvable in the short run. Even if emissions of greenhouse gases are halted altogether, it would take up to a century for the oceans to absorb enough atmospheric carbon dioxide to bring concentrations down to half of preindustrial levels. Nitrous oxide and CFC levels would take at least 50 years to decline by half.

Convincing politicians to take action on global warming is therefore an extremely tricky proposition. As Jacobi puts it: "If you'd taken an MIT class in political science some years ago and said, 'Let's take a problem that human institutions can't deal with,' I think you wouldn't have done better than this. In terms of scale, ambiguity, uncertainty, and distributional implications, it's hard to imagine a more difficult set of circumstances than those presented by global climate warming."

Given the unpredictability of global warming, it is easy to understand why governments have been slow to take precautionary measures. The world's economies are nearly as intricately interconnected and difficult to predict as the global climate. Economic studies of the impact of preparing for global warming read very nearly like the studies of global warming themselves—filled with uncertainties and unintended side effects—and their forecasts can range from worldwide economic disaster to a more rationally ordered world economy. On the one hand, *Forbes* magazine has declared that taking precautions against global warming "could well spell the end of the American dream for us and the world. . . . There is no realistic economic way to prevent a CO_2 doubling without slashing growth and risking a revolt of the have-not nations against the haves." On the other hand, *Worldwatch* magazine, in an article entitled "Saving the Climate Saves Money," insisted that "practical, cost-effective, and quick ways to reduce carbon dioxide emissions abound, and almost without exception they'll bring about an economic boom, not a bust."

Because greenhouse gas emissions are connected to every human activity—from growing rice to raising cattle, from using a refrigerator to driving a car—any policy meant to curb global warming will have to be detailed, dramatic, and international in scope. Proposals to limit global warming may cost billions of dollars to implement, but that may be the price that society has to pay for having mismanaged the environment for so long.

ECONOMIC CONTROLS

When economists try to predict the effect of cutting back on energy consumption and cleaning up the air, they point to the 1970s. The OPEC oil embargoes of 1973 and 1979 accomplished many of the goals proposed by global warming–policy theorists. They raised gasoline prices dramatically, forcing people to consume less energy and leading car manufacturers to create engines that got more miles per gallon. In the United States, the average fuel efficiency of new cars nearly doubled, growing from 14 to more than 26 miles per gallon between 1974 and 1987. Industries dependent on fossil fuels were forced to operate more efficiently, and the federal government invested heavily in solar, wind, and nuclear power. The rate of greenhouse gas emission slowed appreciably during those years. If it had continued to grow at the pre-1973 rate, annual emissions today would be 3 billion tons greater than they are. At the same time, however, the "energy crunch" was accompanied by a particularly nasty economic slump caused by "stagflation." In the United States, both unemployment and inflation rose in what was the worst recession since World War II.

When oil flowed freely again in the 1980s, the nation slowly returned to its wasteful ways. Claims of fuel efficiency gradually disappeared from car advertisements, the budget for alternative energy research in the United States shrank, and laws enforcing energy conservation were relaxed. At the same time, the economy took a turn for the better.

Critics of economic controls on greenhouse emissions point to this pattern as a warning. Any policy that raises the price of energy will hurt the economy. Moreover, some economists argue that creating "clean" technologies that curb greenhouse gas emissions will cost the United States alone as much as $3.6 trillion. According to *Worldwatch* magazine, however, "These studies are simplistic and one-sided, they look only at the total cost of improving energy efficiency or developing non-fossil-based energy resources. The fact that many of these investments may be economical in their own right is completely ignored." The writer goes on to point out that "from 1973 to 1988, the U.S. economy grew 46 percent, while energy use went up only 8 percent. This means the country slashed the

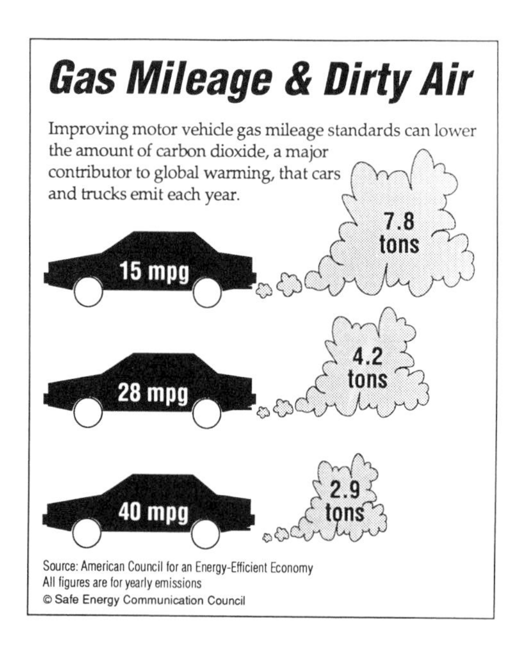

Energy conservation, particularly the reduction of the use of fossil fuels for automobiles, could significantly slow down global warming, not to mention help to reduce air pollution.

amount of energy needed to produce a dollar's worth of goods and services by 26 percent, effectively trimming carbon emissions while strengthening the economy." A paper recently released by the environmental branch of the Dutch government agreed, pointing out that short-term policies aimed at regulating greenhouse gases—taxes on fossil fuels, a global CO_2 budget, and other forms of emission regulation described below—will hardly hamper the economy in the short term (by as little as 3% in the Netherlands, according to one scenario); alternative power research and the development of energy-efficient technologies will stimulate the world economy in the long term.

The most straightforward—and perhaps most effective—method of lowering CO_2 emissions would be to raise the price of fossil fuels by placing a tax on them. The 1970s, once again, are instructive: Gas prices went up, consumption went down. Because the increased cost of fossil fuels would be mandated by the government rather than imposed by market forces, as they were in the 1970s, these taxes could be cleverly structured to encourage people to use the most efficient fossil fuels. Because burning natural gas produces 43% less carbon than does coal to produce the same amount of energy, natural gas would carry a much lower tax. The tax on oil, which produces 18% less carbon than coal, would fall in between. The funds from these taxes could either be given back to consumers in the form of income tax reductions, or they could be used to pay for alternative energy research, reforestation, or any of the other costly programs made necessary by global warming. Clean energy sources, such as solar power and wind power, would have no tax imposed, encouraging their development still further.

An even simpler but perhaps less productive method of limiting greenhouse gases would be for governments to pass laws directly limiting the CO_2 emissions of industries and individuals. Power plants, for instance, would be compelled by law to emit no more than a specified amount of CO_2 per unit of energy generated—in effect, forcing them to switch to cleaner forms of fuel (such as natural gas), pay hefty fines, or offset their emissions through tree planting. Individuals, by the same token, would be allowed to burn only a certain amount of gasoline per year—forcing them to either buy more fuel-efficient cars or rely increasingly on public transportation. According to Worldwatch, redesigning engines to achieve a gas consumption rate of 40 miles per gallon would increase the average cost of a car by $500; the gasoline savings would total $2,000 over the lifetime of the car. If all new cars got 40 miles per gallon, the United States would consume 3 million fewer barrels of oil a day by the year 2005. The EPA has already suggested that cars throughout the world be required to get at least 40 miles per gallon of gasoline and that they be installed with catalytic converters to cut down on ozone and N_2O emissions. It should be noted that car manufacturers have fuel-efficient technology to spare. Toyota has designed a car that can achieve 89 miles per gallon in the city and 110 miles per gallon on the highway. Although such carbon rationing laws can have an immediate impact, they can—when applied too broadly—stifle the economy without paying (as taxes do) for alternatives.

The most innovative method for curbing greenhouse emissions would have to take place on an international rather than a national level. An international environmental agency would determine how much CO_2 could be emitted worldwide in

Bicyclists in London use a separate lane prepared just for them, a common practice in many of the world's larger cities. The bicycle, which uses no fossil fuels and creates no pollution, is the ideal means of transportation for short urban commutes.

a given year. The agency would then divide that amount by the world population, creating a few billion "emission permits" that they would allocate to each country according to population size. Countries would then have to limit CO_2 emissions according to the number of permits received. Of course, this would severely hamper industrialized countries such as the United States and give relatively undeveloped but overpopulated countries more credits than they could use. The industrialized countries, however, could then purchase these surplus CO_2 permits from undeveloped ones, thus helping Third World nations get through their own economic troubles under global warming.

THE THIRD WORLD

The taxes and permits described above are really a sophisticated form of wrist slapping. They are ways to compel relatively rich, industrialized countries to mend their gluttonous ways. Together, the United States, the Soviet Union, Japan, and Germany produce nearly 50% of the world's greenhouse emissions, though their combined populations are less than 25% of the world total. The United States emits 4.2 tons of carbon for each of its citizens, more than twice the amount emitted by Japan. Clearly, the burden of reform lies with the developed countries.

Unfortunately, the developed nations alone cannot stop global warming. By 2025, the United Nations Populations Division projects that the world population will grow by 3.2 billion people. Fully 3 billion of these new souls will be born in Third World nations—leaving the First World with only 16% of the total world population. Although the Third World currently only accounts for 40% of global greenhouse gas emissions, that percentage should rise to 50% or more by the year 2025 and continue to spiral upward thereafter. If three-quarters of the world's underdeveloped nations industrialize in the near future, they could triple the rate of greenhouse gas emissions.

Most developing nations will not be able to afford the expensive "clean" technologies that the First World will develop over the next few decades. They will turn primarily to coal—the dirtiest of fossil fuels—and oil to fuel their power plants. By the same token, the cars sold in the Third World—which will grow in number along with the population and industrialization—rarely have catalytic converters or gas tanks designed for unleaded gasoline. As populations grow, more Third World forests will be cleared to make room for farms and cities. Finally, Third World countries will be too poor to prepare for rising sea levels, more frequent floods, and erratic water supplies—making them most vulnerable to the effects of global warming.

In order to control burgeoning greenhouse gas emissions without robbing the Third World of its right to industrialize, developed countries may have to lend a strong helping hand. The First World could subsidize the development of alternative energy sources in developing countries, pardon debts in the Third World in exchange for limits on deforestation, or—in the event that a system of emissions permits is developed—exchange those permits for further limits on deforestation. To prepare the Third World for destructive floods and global warming's other side effects, developing countries will have to receive more grants for the building of seawalls, stronger dams, new reservoirs, and other safety measures. The final statement of the Second World Climate Conference, held in Geneva in October 1990, with representatives from 130 countries, offered strong hope that such measures were forthcoming. It recommended that "adequate and additional financial resources" and the "best available environmentally sound technologies" be given to Third World countries on a "most favorable basis."

AGREEMENTS TO DATE

Global warming is a uniquely international problem. Greenhouse gases do not respect national borders. If Thailand releases 48 million metric tons of CO_2 into the air every year, then the entire world pays for it. By the same token, if the Netherlands halts its greenhouse emissions entirely, but the rest of the world continues polluting as before, then the Dutch effort comes to naught. For this reason, curbing global warming is both an immensely complex political problem and a chance to test the waters of world government. Scientists are increasingly finding that very few environmental problems are isolated or self-contained. Acid rain, water pollution, pesticide contamination, nuclear waste disposal, and deforestation all have far-reaching effects on the biosphere. Imposing global limits on CO_2 emissions or distributing emissions permits requires countries to cooperate as never before.

One of the first major conferences to call for emissions restrictions took place in Toronto in June 1988 and included 300 officials from 46 countries. The conference called for CO_2 emissions to be reduced "by approximately 20% of the 1988 levels by the year 2005 as an initial global goal." Unfortunately, when this proposal was submitted to an international environmental summit meeting in November 1989, the United States, Japan, and the Soviet Union vetoed it, pleading the need for more scientific and economic studies. By the time of the Second World Climate Conference, the global consensus had grown even more firm on the need for CO_2 limits. Japan had agreed to stabilize its CO_2 emissions at 1990 levels by the year 2000. Australia had agreed to reduce emissions of CO_2 and other greenhouse gases by 20% before 2005, and the Netherlands had agreed to stabilize emissions at 1990 levels by 1995 and to reduce emissions by a further 3% to 5% by 2000. Germany, with perhaps the most ambitious plan, hopes to cut CO_2 emissions by

25% by the year 2005, without curbing economic growth in the process. The European Community as a whole has agreed to freeze CO_2 emissions at 1990 levels by the year 2000. Nevertheless, the conference failed to establish a worldwide freeze, chiefly because of objections from the United States, the Soviet Union, and Saudi Arabia. A world climate treaty is scheduled to be negotiated in February 1992—by which time, those in favor of emissions limits believe, the global consensus favoring such limits will have grown strong enough to sink any objections from the United States. The state of Oregon, in any case, has already jumped ship. It has passed a state law requiring a 20% decrease in greenhouse gas emissions by the year 2005.

Fickle as it may be, public opinion is the key to maintaining a global consensus on the need for emissions limits. Global warming has had a good run in the media for the last two years. But as Charles Keeling, the first scientist to document the steady rise of CO_2 in the atmosphere, has pointed out, a few cool years in a row—part of the climate's natural fluctuations—would probably convince most people that global warming is a hoax. To prevent such complacency and to maintain pressure on governments to curb greenhouse gas emissions, people must continue to educate themselves about the sources and long-term effects of CO_2 and other chemical compounds. Global warming is one problem that will not go away with yesterday's news.

A technician inspects one of the turbines at the Eklutna hydroelectric power plant in Alaska. Although hydropower has its problems, it is a clean way to produce energy, and the EPA believes that the United States could produce 50% more electricity just by fitting existing dams with these turbines.

chapter 8

PREPARING FOR THE FUTURE

Cleaning up the air, preserving the forests, learning to conserve energy—the measures needed to curb global warming—make basic environmental sense. In some cases, declining greenhouse gas levels will be the fringe benefits of other efforts to clean up the environment. Plans to rid Los Angeles of smog are expected to lower the city's greenhouse gas emissions. Saving or replanting the rainforest to preserve tropical plant and animal species will cut down on carbon dioxide–producing fires and allow more CO_2 to be absorbed through photosynthesis. Reducing the world's dependence on fossil fuels will limit greenhouse gases rising from car exhausts and industrial smokestacks in addition to preparing society for the days when fossil fuels have been exhausted for good.

A KILOWATT SAVED . . .

Conspicuous consumption has long been one of the First World's greatest vices. These societies must learn to conserve resources, particularly energy resources. In many cases, energy-efficient technology is already available—individuals and industries simply have not felt the need to adopt it. Yet simple efficiency is the most powerful tool available for combating global warming. According to one 1987 report, industrial countries could consume 2% less fossil fuel every year just with existing technology. The Worldwatch Institute estimates that current technology can make electric motors 15% to 40% more efficient. Cogeneration plants—power plants that produce both electricity and heat—could reduce CO_2 emissions by 25% to 40%.

Energywise, American houses are filled with holes. The most efficient homes currently built use 30% less energy per square foot than the average house in the United States. Studies show that houses combining all the latest designs and technology could use 90% less energy. Refrigerators, the biggest electricity gobblers in most houses, could be made 75% more efficient. In the United States, a new law requires that refrigerators be 25% more efficient on average by 1992. More efficient lighting technology, coupled with well-designed buildings, can cut the energy needed for lighting by 75%. Simply doubling the efficiency of electric lighting could reduce projected carbon emissions worldwide by 225 million tons by the year 2010. Overall, the Worldwatch Institute estimates that improving energy efficiency by 3% every year until the year 2010 could keep 3 billion tons of carbon from reaching the atmosphere.

ALTERNATIVE ENERGY SOURCES

Fossil fuels currently provide 78% of the world's energy, but such fuels are being consumed approximately 10 million times faster than nature creates them. As these resources rapidly disappear, the best way to keep the world's industry from grinding to a halt is to find new sources of energy—preferably non-polluting ones. Wind energy, solar energy, geothermal energy, hydropower, nuclear energy, and biomass energy can turn engines just as effectively as fossil fuels. Although efforts to develop such alternative energy sources stalled during the 1980s, these sources will be the key to curbing and then reversing global warming in the 21st century, but much research remains to be done and many problems must be solved.

Nuclear power is a case in point. Once viewed with great optimism, nuclear power has fallen on hard times. Although nuclear plants today provide 15% of the world's electricity, the number of plants is dwindling rather than increasing. The growth of nuclear power has slowed from 7% annually during the 1970s to 1.8% during the 1980s. Once projected to reach a 4-million-megawatt capacity worldwide by the year 2000, nuclear energy will only produce 400,000 megawatts by that time. According to Worldwatch, even if the 94 nuclear power plants that were under construction worldwide in 1989 are completed—an unlikely event, given the level of public protest against nuclear power in recent years—they will prevent only an additional 110 million tons of carbon from being emitted into the atmosphere from fossil fuels—just about the amount of carbon emitted by Italy alone. Nuclear power has quite simply proved too costly, too unreliable, and too dangerous. Sites where nuclear wastes can be safely stored still have not been found, and the Three Mile Island and Chernobyl disasters have severely shaken public trust. In the United States, polls show that people oppose nuclear power two to one, and more countries have rejected nuclear power over the last few years than have accepted it.

A hope for the future, this doughnut-shaped experimental fusion reactor, sometimes called a stellarator or a tokamak, might produce large amounts of clean energy from the hydrogen atoms in sea water, but the heat that many such reactors would generate might aggravate global warming.

Advocates of nuclear power argue that the design flaws that caused past disasters have been largely eliminated. New plants could be built, they say, that are "inherently safe." Designed in separate modules and cooled with helium gas rather than water, such plants can avoid a meltdown even if the cooling system fails and no safety precautions are taken. Although such nuclear plants are a great deal safer than conventional reactors, they produce much less energy per unit of fuel, mainly because their safety is a consequence of operating at lower temperatures. The need for safety is gradually pricing nuclear power out of the market. A kilowatt-hour of electricity produced in a nuclear power plant costs up to 15 cents—more than double the cost of coal-generated power and more than many of the cleaner technologies described below.

Harnessing the sun's energy has been the dream of engineers for the last 40 years. During the energy crisis of the 1970s, government funding spurred researchers to design increasingly efficient solar energy systems. Consumers, discouraged by high oil prices and encouraged by federal tax credits for installing solar collectors, found the new systems more affordable. When oil prices dropped in the 1980s, however, the market for solar collectors disappeared, along with federal tax credits and federal funding for solar research. Sales of solar energy systems dropped 70%, and 14 out of every 15 people in the industry lost their job.

When solar power research dried up, the existing technology had not yet achieved a high level of efficiency. Photovoltaic cells had been developed that could convert sunlight directly into electricity using semiconductor materials, but the average cell converts only 12% of the sunlight that strikes it into electricity, at an optimum cost of 30 cents per kilowatt-hour—5 times the cost of energy from a new coal-fired plant. An innovative solar thermal system built by the LUZ Corporation in the Mojave Desert uses sunlight to heat oil-filled tubes. The oil then turns water into steam to power turbines that generate electricity. This system converts 22% of the sunlight that strikes it into electricity, at a cost competitive with coal-generated electricity. Ultimately, however, solar-thermal power will be only a stopgap measure while photovoltaic technology is perfected.

As global warming and higher oil prices loom on the horizon, interest in solar power is again growing. The market for solar power in the United States grew by an estimated 50% between 1988 and 1989, thanks in large part to solar-powered home electrical systems, water pumps, calculators, and other small-scale devices. As research increases, the coming decades should see dramatic progress in solar-power technology. Cheap materials, such as amorphous silicon, could drastically cut the cost of photovoltaic cells; materials such as gallium arsenide could triple the cells' efficiency. Over the last decade, the cost of photovoltaic cells dropped by 90%. Richard Swanson, director of SunPower, predicts that his company will be able to build arrays

The collectors of this solar-thermal system, built by the LUZ Corporation in the Mohave Desert, focus sunlight on a tube of oil. The heated oil turns water into steam, and the steam pushes turbines to generate electricity, without any atmospheric pollution or emission of greenhouse gas.

of photovoltaic cells within the next four years that can generate electricity at eight cents per kilowatt-hour, less than a third of the cost of previous systems. Such inexpensive systems will be of great value to Third World countries, where the high cost of laying electrical lines from conventional power plants to remote areas has made even inefficient photovoltaic cells cost-effective. According to Worldwatch, India alone has installed 6,000 solar energy systems in small villages.

As scientists perfect superconducting power cables that carry electricity extremely efficiently, large cities will be able to draw solar energy from photovoltaic cells arrayed in distant, sun-drenched areas. The countries of Europe are already looking at plans to build huge solar-powered generators in North Africa. According to the Solar Energy Research Institute, within the next 40 or 50 years, photovoltaic cells could produce more than half the electricity in the United States.

Wind power also has great unexplored potential. First used to generate electricity in the 1890s, wind was responsible for a quarter of Denmark's energy production by the early 20th century. Although it was displaced by fossil fuels for most of the 20th century, wind power made a comeback during the 1970s. As with solar energy, however, funding for crucial research dropped precipitously during the 1980s—from $86 million in 1981 to $30 million during the last few years. Nevertheless, wind power is so affordable, with average costs of six to eight cents per kilowatt-hour, that wind-powered generators have begun to sprout up all over the world. There are 20,000 wind turbines in the world today. They already provide for 15% of San Francisco's energy needs and a growing percentage of the energy needs of China and India. It is estimated that wind power could provide one-quarter of all U.S. electricity by the beginning of the 21st century and about 10% of the world's electricity by the year 2050. New production techniques could lower the price of wind power by 25% in the next 5 years.

Volcanoes, geysers, and hot springs are another potential source of energy. Today *geothermal* power is a fast-growing industry, with 150 geothermal power plants around the world already producing energy at 4 to 8 cents per kilowatt-hour. On the slopes of dormant volcanoes in Iceland, crops are grown in geothermally heated greenhouses, and 65% of all houses are kept warm by water from underground hot springs. Depending on how actively the United States pursues geothermal power, it could provide anywhere from 4,000 to 19,000 megawatts of power by the year 2000—only a fraction of the country's energy needs, but a beginning nonetheless.

Hydroelectric power has a longer history. By the early 20th century, dams fitted with turbines were generating 60% of total electricity output in the United States. Although fossil fuels now provide most of the world's energy supply, hydroelectric plants still produce 25% of the world's electricity, and the number of plants continues to grow. According to the World Bank, 31 developing countries doubled their hydropower capacity between 1980 and 1985. Pakistan relies on water rushing down from the Himalayas for 65% of its electrical needs, and Nepal has only just begun to tap its hydroelectric potential. Brazil, with the largest dams in the world, looks to hydropower for more than 90% of its electricity, as does the Northwest United States. Although hydropower produces no greenhouse gases, it also has its share of unpleasant side effects. Villages, vast tracts of forests, and rare habitats are often destroyed when dams force

Located on the island of Oahu in Hawaii, this wind turbine is the largest and most powerful in the world. It is 360 feet high and generates more than 8 million kilowatt-hours of electricity every year, without adding any greenhouse gases to the atmosphere.

rivers to flood their banks, and many environmentalists mourn the loss of the world's free-moving waterways. Moreover, if dams are not properly designed, they may be vulnerable to the catastrophic floods that will occur more frequently in a warmer world.

Instead of building new dams, many countries are putting their existing dams to work. Most of the world's 60,000 dams have never been fitted with turbines. The EPA estimates that unutilized dams in the United States could provide the nation with nearly 50% more electricity if they were fitted with turbines.

Biomass energy, that is, energy from organic waste, is an application as ancient as civilization itself. Locked within the garbage that humans throw away every day—peanut shells, withered wheat, paper cartons, livestock dung, wood chips, apricot pits, corncobs, rice hulls—is literally energy to burn. The Third World still relies on such age-old sources of fuel for 50% of its energy needs. In China, more than 8 million households contain "biogas digesters" that convert manure and other wastes into methane—which is then used for cooking and lighting. In India, on the other hand, 575 million pounds (259 million kilograms) of dung are ignored each year while people continue to burn wood for fuel. Even in areas where fuel wood is scarce, open fires are used to cook meals when inexpensive wood stoves could provide the same heat with a fraction of the fuel.

There are many efficient methods of turning biomass into fuel or energy. Many are already economically competitive. Power plants run off the methane fumes rising from landfills, removing this potent greenhouse gas from the atmosphere in the process. Various industries in the United States have shown that power plants can run cheaply on wood chips, peach and cherry pits, or peanut shells. Corncobs, sugarcane, cassava, and rice hulls can be converted into ethanol to fuel cars and buses. Ethanol, when burned, emits far less CO_2 than does gasoline. Altogether, such methods could supply up to 30% of the world's energy needs by the year 2050.

REFORESTATION

Of all the efforts to control global warming, reforestation has struck the most responsive chord with the public. The idea is simple and persuasive: If burning the world's rainforests sends CO_2 into the atmosphere, then replanting them will pull that carbon back into living plant tissue. Compared to the amount of carbon stored in living things, the carbon released through the burning of fossil fuel is minuscule. Every hectare of tropical forest can hold about 5.5 tons of carbon. If marginal croplands, abandoned pastures, and other degraded soils around the world were planted with trees, they could absorb millions of tons of carbon every year.

Such statistics and the natural appeal of trees have spawned dozens of tree-planting proposals over the last few years. Australia has announced a program to plant 1 billion trees by the year 2000, and in the United States the Bush administration has launched what it calls "a major reforestation initiative" called America the Beautiful. American citizens, sponsored by federal and private funds, will plant 1 billion trees every year until a total of 10 billion have been planted. One Connecticut power company that employs a coal-fired plant decided to atone for spewing 387,000 tons of carbon into the air annually by sponsoring the planting of 52 million trees in Guatemala. In the United States and Europe, dozens of groups are organizing urban tree-planting campaigns. In Los Angeles, a group called TreePeople is trying to plant several million trees over the next few years.

Such programs also cut down on soil erosion and expand wildlife habitats in the countryside. They beautify neighborhoods and provide shade in the city, cutting down on the need for air-conditioning, a prime cause of greenhouse gas emissions. But they will hardly stop or reverse global warming. To do so, each and every person on earth would have to plant and maintain 1,000 trees in a combined area as large as Europe. Even if 5 trillion trees could be planted, they might not have the desired effect. Much of the bare land that would be planted has a high albedo, meaning that it reflects a good deal of sunlight back into space. Once covered with forest, this land would absorb 20% more sunlight than before, raising global temperatures rather than lowering them.

The best way to keep carbon out of the air is to first preserve the existing forests. Every hour, 500,000 trees are destroyed in tropical forests. In order to slow such deforestation, the nations of the First World must find ways for Third World countries to escape their often crushing international debts and help them build economies strong enough to survive without revenues from logging and cattle ranching.

INDIVIDUAL INITIATIVE

The drama of global warming is played out on such a huge scale that many people feel helpless to combat it. The earth's climate, the march of industry, the population boom, society's endless appetite for energy—the forces at work are so large and so complicated that only whole nations, signing global treaties, seem capable of dealing with them. Yet many of the problems described in these chapters originate in a few simple activities performed by ordinary citizens throughout the industrial world: driving a car, eating a thick steak, using a can of hair spray, turning on the air conditioner, and so on.

The following are suggestions for life under the threat of global warming. They hardly require a radical change of life-style, just a sense of responsibility toward future generations.

- Walk, ride a bike, or buy a car that gets more than 40 miles per gallon of gas.

- Conserve energy at home. Turn off lights in unoccupied rooms; install fluorescent lights and an energy-efficient refrigerator. Wear a sweater in the winter and sweat a little in the summer.
- Consider installing a solar water heater or other devices that do not depend on fossil fuels.
- Recycle aluminum, plastic, and paper to prevent needless mining, pollution, and deforestation.
- Eat less meat. Livestock produce methane at an astonishing rate. As the demand for beef decreases, so will the need to cut down tropical forests for cattle pastures.
- Avoid using products that contain chlorofluocarbons—Styrofoam, spray cans, and air conditioners using Freon.

And most of all, do not be fooled when the weather turns cooler.

APPENDIX: FOR MORE INFORMATION

Environmental Organizations

American Council for an Energy Efficient Economy
1001 Connecticut Avenue NW
Suite 535
Washington, DC 20036
(202) 429-8873

Climatic Institute
316 Pennsylvania Avenue SE
Washington, DC 20003
(202) 547-0104

Energy Conservation Coalition
1525 New Hampshire Avenue NW
Washington, DC 20036
(202) 745-4874

Environmental Action Coalition
625 Broadway
New York, NY 10012
(212) 677-1601

Environmental Action Foundation
1525 New Hampshire Avenue NW
Washington, DC 20036
(202) 745-4879

Environmental Defense Fund
257 Park Avenue South
New York, NY 10010
(212) 505-2100

Friends of the Earth
530 Seventh Street SE
Washington, DC 20003
(202) 543-4312

Global Greenhouse Network
1130 17th Street NW
Washington, DC 20036
(202) 466-2823

Global Releaf
c/o the American Forestry Association
P.O. Box 2000
Washington, DC 20013
(202) 667-3300

Greenhouse Crisis Foundation
1130 17th Street NW/630
Washington, DC 20036
(202) 466-2823

Greenpeace USA
1436 U Street NW
Washington, DC 20009
(202) 462-1177

New Alchemy Institute
237 Hatchville Road
East Falmouth, MA 02536
(508) 564-6301

Rainforest Action Network
301 Broadway
San Francisco, CA 94133
(415) 398-4404

United Nations Environment Programme
New York Liaison Office, Room DC2-0803
United Nations
New York, NY 10017
(212) 963-8093

World Resources Institute
1709 New York Avenue NW
Washington, DC 20036
(202) 638-6300

Worldwatch Institute
1776 Massachusetts Avenue NW
Washington, DC 20036
(202) 452-1999

Government Agencies

U.S. Council for Energy Awareness
P.O. Box 66103
Dept. AY31
Washington, DC 20035
(202) 293-0770

Environmental Protection Agency (EPA)
401 M Street SW
Washington, DC 20460
(202) 382-2090

PICTURE CREDITS

Courtesy of Agricultural Research Service, USDA: p. 47; AP/Wide World Photos: pp. 17, 18, 22, 49, 52, 54, 62, 82, 89; Delaney/EPA: p. 44; Department of Energy: pp. 94, 98, 103; Greenpeace/Larry Lipsky: p. 68; Greenpeace/Richard Lord: p. 37; Greenpeace/Marriner: p. 40; Greenpeace/Midgley: p. 78; John Kauffmann/National Park Service: p. 59; Luz International Limited: p. 100; NASA: pp. 12, 29, 30, 32; NASA/Goddard Institute for Space Studies: pp. 20, 73; National Center for Atmospheric Research: pp. 70, 75; NOAA: pp. 42, 64; Safe Energy Communication Council: p. 86; USDA photo: p. 56

FURTHER READING

Abrahamson, Dean E., ed. *The Challenge of Global Warming.* Covelo, CA: Island Press, 1989.

Begley, Sharon. "Heat Waves." *Newsweek* (July 11, 1988).

Brookes, Warren T. "Global Warming Panic: A Classic Case of Overreaction." *Forbes* (December 25, 1989).

Firor, John. *The Changing Atmosphere.* New Haven: Yale University Press, 1990.

Fisher, David E. *Fire and Ice.* New York: HarperCollins, 1990.

Flavin, Christopher. *Worldwatch Paper 91—Slowing Global Warming: A Worldwide Strategy.* Washington, DC: Worldwatch Institute, 1989.

Flavin, Christopher, and Nicholas Lenssen. "Saving the Climate Saves Money." *Worldwatch* (December 1990).

Graedel, Thomas E., and Paul J. Crutzen. "The Changing Atmosphere." *Scientific American* (September 1989).

Hassol, Susan, and Beth Richman. *Energy: 101 Practical Tips for Home and Work.* Snowmass, CO: Windstar Foundation, 1989.

Houghton, J. T., J. J. Ephraums, and G. J. Jenkins, eds. *Climate Change: The IPCC Scientific Assessment.* Cambridge: Cambridge University Press, 1990.

Houghton, Richard A., and George M. Woodwell. "Global Climatic Change." *Scientific American* 260 (no. 4, 1989): 36–44.

Jacobson, Jodi. "Swept Away: Rising Waters and Global Warming." *Worldwatch* (Jan./Feb. 1989): 20–26.

Lean, Geoffrey, Don Hinrichsen, and Adam Markham. *The World Wildlife Fund Atlas of the Environment.* New York: Prentice- Hall, 1990.

Lester, R. T., and J. P. Myers. "Global Warming, Climate Disruption, and Biological Diversity." In *Audubon Wildlife Report 1989/1990.* New York: Academic Press, 1990.

McKibben, Bill. *The End of Nature.* New York: Random House, 1989.

Mintzer, Irving M. *A Matter of Degrees: The Potential for Controlling the Greenhouse Effect.* Washington, DC: World Resources Institute, 1987.

National Academy of Sciences. *Ozone Depletion, Greenhouse Gases, and Climate Change.* Washington, DC: National Academy Press, 1989.

Oppenheimer, Michael, and Robert H. Boyle. *Dead Heat: The Race Against the Greenhouse Effect.* New York: Basic Books, 1990.

Schneider, Stephen H. "The Changing Climate." *Scientific American* (September 1989): 70–79.

———. *Global Warming: Are We Entering the Greenhouse Century?* San Francisco: Sierra Club Books, 1989.

Shabecoff, Philip. "Major Greenhouse Impact Is Unavoidable, Experts Say." *New York Times* (July 19, 1988): p. C1.

Steger, Will, and Jon Bowermaster. *Saving the Earth: A Citizen's Guide to Environmental Action.* New York: Knopf, 1990.

Waggoner, Paul E., ed. *Climate Change and U.S. Water Resources.* New York: Wiley, 1990.

Wiener, Jonathan. *The Next One Hundred Years: Shaping the Fate of Our Living Earth.* New York: Bantam Books, 1990.

Wilson, E. O., ed. *Biodiversity.* Washington, DC: National Academy Press, 1988.

World Resources 1990–91: A Guide to the Global Environment. New York: Oxford University Press, 1990.

Young, Louise B. *Sowing the Wind: Reflections on the Earth's Atmosphere.* New York: Prentice-Hall, 1990.

Conversion Table

(From U.S./English system units to metric system units)

Length

1 inch = 2.54 centimeters
1 foot = 0.305 meters
1 yard = 0.91 meters
1 statute mile = 1.6 kilometers (km.)

Area

1 square yard = 0.84 square meters
1 acre = 0.405 hectares
1 square mile = 2.59 square km.

Liquid Measure

1 fluid ounce = 0.03 liters
1 pint (U.S.) = 0.47 liters
1 quart (U.S.) = 0.95 liters
1 gallon (U.S.) = 3.78 liters

Weight and Mass

1 ounce = 28.35 grams
1 pound = 0.45 kilograms
1 ton = 0.91 metric tons

Temperature

1 degree Fahrenheit = 0.56 degrees Celsius or Centigrade, but to convert from actual Fahrenheit scale measurements to Celsius, subtract 32 from the Fahrenheit reading, multiply the result by 5, and then divide by 9. For example, to convert 212° F to Celsius:

212 – 32 = 180 x 5 = 900 ÷ 9 = 100° C

GLOSSARY

biomass energy Energy from organic waste. Combustible fuels, such as **methane** and alcohol, may one day meet a large portion of the world's energy needs.

biosphere All life that inhabits the earth—described as a biosphere because living things effectively surround the planet much as the atmosphere does.

carbon dioxide (CO_2) A colorless, odorless, incombustible gas present in the atmosphere, the molecules of which are composed of one atom of carbon and two atoms of oxygen.

catalytic converter A device installed on the exhaust system of an automobile to convert pollutants such as carbon monoxide into less harmful gases.

chlorofluorocarbons Chemical compounds developed in the 1930s as refrigerants and later used in spray cans and other products. CFCs, as they are called, are both harmful to the ozone layer and are important greenhouse agents.

climatology The scientific study of the earth's climate.

cyclone A rotating storm system that is often accompanied by steady high winds and torrential rains; cyclones may become much more frequent in an era of rising global temperatures.

deforestation The destruction of any forest by human beings—most often used in reference to the current situation in the tropics, where rainforests are being leveled by the millions of acres yearly. The planting of forests to offset this trend is called reforestation.

ethanol A liquid fuel made from **biomass**. Burning ethanol produces fewer **greenhouse gases** than does burning gasoline and other **fossil**

fuels; this is one reason that ethanol is becoming an increasingly attractive alternative to gasoline.

fossil fuel Any combustible organic material, such as oil, coal, or natural gas, derived from the remains of ancient life.

general circulation model (GCM) Complex computer programs that simulate the earth's climate, allowing atmospheric scientists to predict future climate patterns.

greenhouse effect The trapping of infrared radiation in the earth's atmosphere by gases such as **carbon dioxide (CO_2)** and methane (CH_4), which results in higher temperatures.

greenhouse gases Gases such as carbon dioxide, methane, **ozone**, **nitrous oxide**, **chlorofluorocarbons**, and **water vapor** that, when in the atmosphere, block heat energy rising from the earth's surface and radiate it back earthward. These gases are the linchpins of the **greenhouse effect**.

kilowatt-hour A unit of energy equivalent to the energy transferred or expended in 1 hour by 1 kilowatt of power; approximately 1.34 horsepower-hours.

methane (CH_4) A gas released as organic matter decomposes. Methane is one of the most potent greenhouse gases, and many scientists foresee a dramatic increase in the amount of methane released as the earth warms.

metric ton A unit of 1,000 kilograms, equivalent to 2204.62 pounds.

nitrous oxide (N_2O) A gas released into the atmosphere by coal and forest fires, car exhaust, and nitrogen-based artificial fertilizers. Popularly known as laughing gas, nitrous oxide is a powerful greenhouse agent.

old growth forest A forest that has never, within known history, been cut down.

paleoclimatology The study of ancient climates.

ozone (O_3) Ozone is much talked about as the building block of the ozone layer, which protects life on earth by blocking out destructive ultraviolet radiation from the sun. However, ozone also exists in the lower atmosphere in increasingly high concentration, where it acts as a greenhouse agent.

paleoclimatologists Historians of the earth's climate.

photosynthesis A process, powered by sunlight, by which plants draw in carbon dioxide, water, and inorganic salts, and turn them into oxygen and the carbohydrates that allow them to grow.

respiration The reverse of **photosynthesis**, by which plants expel carbon dioxide back into the atmosphere at night. As the earth warms, plants will likely respire CO_2 into the atmosphere at a far greater rate than they produce oxygen and thus accelerate global warming.

troposphere The lowest region of the atmosphere, in which most weather takes place; extends from the surface of the earth up to an average altitude of 7.5 miles (12 kilometers).

tsunami A tidal wave. Enormously destructive when they strike a coastal city, tsunamis will become more frequent as the earth warms.

water vapor Water in gaseous form, the result of evaporation of water from the earth's lakes, seas, and oceans. As the surface temperature of these bodies of water rises, more evaporation will occur, and more water vapor will enter the atmosphere. When water vapor condenses in the cooler air at higher levels, clouds form. Higher global temperatures, therefore, may increase the earth's cloud cover. It is not yet known how this process will affect global warming.

INDEX

ABOUT THE AUTHOR

BURKHARD BILGER is a journalist and a writer. Raised in Oklahoma, he traveled to Yale University to earn his B.A. in English. His articles have appeared in the *Boston Globe*, the *Boston Phoenix*, the *Brooklyn Phoenix*, *Location Update*, and *Oklahoma Today*. He has worked as a researcher for the *Nation* and is currently an associate editor for *Earthwatch* magazine. Bilger's work for *Earthwatch* has taken him to Tunisia, Majorca, Poland, Switzerland, South Africa, Namibia, and western Massachussetts to cover stories on subjects ranging from pollution in Poland to the politics of African archaeology.

ABOUT THE EDITOR

RUSSELL E. TRAIN, currently chairman of the board of directors of the World Wildlife Fund and The Conservation Foundation, has had a long and distinguished career of government service under three presidents. In 1957 President Eisenhower appointed him a judge of the United States Tax Court. He served Lyndon Johnson on the National Water Commission. Under Richard Nixon he became under secretary of the Interior and, in 1970, first chairman of the Council on Environmental Quality. From 1973 to 1977 he served as administrator of the Environmental Protection Agency. Train is also a trustee or director of the African Wildlife Foundation; the Alliance to Save Energy; the American Conservation Association; Citizens for Ocean Law; Clean Sites, Inc.; the Elizabeth Haub Foundation; the King Mahendra Trust for Nature Conservation (Nepal); Resources for the Future; the Rockefeller Brothers Fund; the Scientists' Institute for Public Information; the World Resources Institute; and Union Carbide and Applied Energy Services, Inc. Train is a graduate of Princeton and Columbia Universities, a veteran of World War II, and currently resides in the District of Columbia.